MATRIX LOGIC

MATRIX LOGIC

August Stern

1988

NORTH-HOLLAND
AMSTERDAM · NEW YORK · OXFORD · TOKYO

ISBN: 0 444 70432 9

Publishers:
Elsevier Science Publishers B.V.
P.O. Box 1991
1000 BZ Amsterdam
The Netherlands

Sole distributors for the U.S.A. and Canada:
Elsevier Science Publishing Company, Inc.
52 Vanderbilt Avenue
New York, N.Y. 10017
U.S.A.

LIBRARY OF CONGRESS
Library of Congress Cataloging-in-Publication Data

Stern, August, 1945-
Matrix logic / August Stern.
p. cm.
Includes index.
ISBN 0-444-70432-9
1. Matrix logic. I. Title.
QA9.9.S74 1988
511.3--dc19
88-11039
CIP

PRINTED IN THE NETHERLANDS

Acknowledgements

Matrix logic is formulated across the sciences, from logic and mathematics to physics and computer science. Considerable effort is required to achieve a complete and yet coherent presentation.

The author wishes to thank Prof. K. Siegbahn for useful advice given in the initial stages of this work, and particularly Prof. J.L. Mackenzie for devoting much time to improving the quality of the text.

The author would be most appreciative of suggestions for further improvements.

August Stern, Amsterdam

CONTENTS

INTRODUCTION

The development of symbolic logic follows in the footsteps of mathematics and physics. Formulated in the middle of the nineteenth century by George Boole, symbolic logic was initially used to clarify the difficulties of Aristotelian logic. It has proved extremely useful in the domain of mathematical reasoning, and together with set and group theories today plays an important role in the foundations of mathematics. The fate of symbolic logic has been greatly affected by the development of modern physics, and especially by the computer age, where it has become an indispensable instrument of computer design and programming.

The use of symbolic notation, which characterizes a mathematical method, has brought logic closer to the realm of mathematics. However, in spite of major advances of logic achieved over the past hundred years and extensive studies of its relations with algebras, number theory and mathematical foundations of computer science, symbolic logic has not yet attained a genuinely computational status. Mathematics cannot be properly dealt with without techniques that afford some rigorous means of counting, and the lack of such overall counting machinery is precisely what separates the domain of logic from that of mathematics. Though the boundary between logic and mathematics is somewhat vague and overlapping, it is nevertheless real and significant.

The purpose of this book is to extend the computational capabilities of logic, and thus to take a step towards closing the gap between the domains of logic and mathematics. In the course of forging a link between formal logic and vector space formalism, we develop a matrix formulation of logic.

The development of matrix logic is particularly important since the ideas involved in the theory of vector spaces are of fundamental significance in much of modern mathematics.

The newly discovered links between logic and matrix algebra lead to a coherent theory, in which the power of direct mathematical computation is applied to the domain of logic operations. Matrix logic catalyzes the convergence of several disparate areas which previously were considered not to be directly related.

We begin the development of matrix logic by interpreting the Boolean logic connectives as matrix operators acting in the two adjoining spaces of logic vectors. This allows us to incorporate the set of 20 Boolean operations into one universal device of matrix multiplication. With the operations of conventional logic becoming genuinely mathematical operations, we extend logic into the domains of negative, continuous, and multi-dimensional truth-values.

An investigation of the intricate properties of the newly defined logic operators leads to the interesting parallels between logic operators and quantum-mechanical operators in the Dirac and Heisenberg formulations of quantum mechanics. What is seen from the logical point of view as a decision inference process can be described in quantum language as the act of measurement. Considering the general scheme for translating between the two interpretations, we propose to view logic not as an abstract construct but as a fundamental structure lying at the base of real physical interactions. This allows us to interpret logic as a macroanalogue of spin and to address logic valuations in terms of the space-time diagram method of quantum field theory.

The matrix formulation of logic creates a number of attractive possibilities in computer science. Matrix logic is not just a new comfortable notation, but, through the application of the very powerful mathematical apparatus of matrix calculus, it offers the possibility of exploring and interpreting the rich results already obtained from the point of view of logic and vice versa. With the matrix formulation come effective computational procedures that are based on entirely new and very powerful methods of logical inference. The further development of matrix logic promises to influence research into artificial intelligence and the intelligence code.

Applicable in a wide variety of fields, matrix logic demonstrates a large potential, making it a useful tool with broad spectrum of theoretical and technical applications in computer science, fundamental physics, as well as in mathematics and logic themselves.

In view of the multidisciplinary nature of matrix logic applications, every effort has been made to provide a comprehensive and explicit text, accessible to anyone with a basic mathematical background. In this book the reader will find an elementary yet complete introduction to the basic principles of matrix logic, and its most important results and applications.

In order to allow this book to be used as a text-book, and to make the material accessible to engineers seeking immediate utilization, we avoid wherever possible complicated mathematics and theorem proving, instead placing the main emphasis on the computational properties of matrix logic. Though consultation of related sources on matrix algebra might be useful, we have included all the necessary introductory material, thus providing a complete source for the study of matrix logic, so that this book is complete in itself.

As a result it is expected that the book can be read by a logician who is not familiar with matrix calculus, by a physicist who has only general knowledge of symbolic logic, and by a computer scientist who may not be at ease with quantum physics. The book is divided into two overlapping parts: the first is devoted to the theoretical aspects of matrix logic, and the second to its applications. Schematically the organization of the book can be presented as follows:

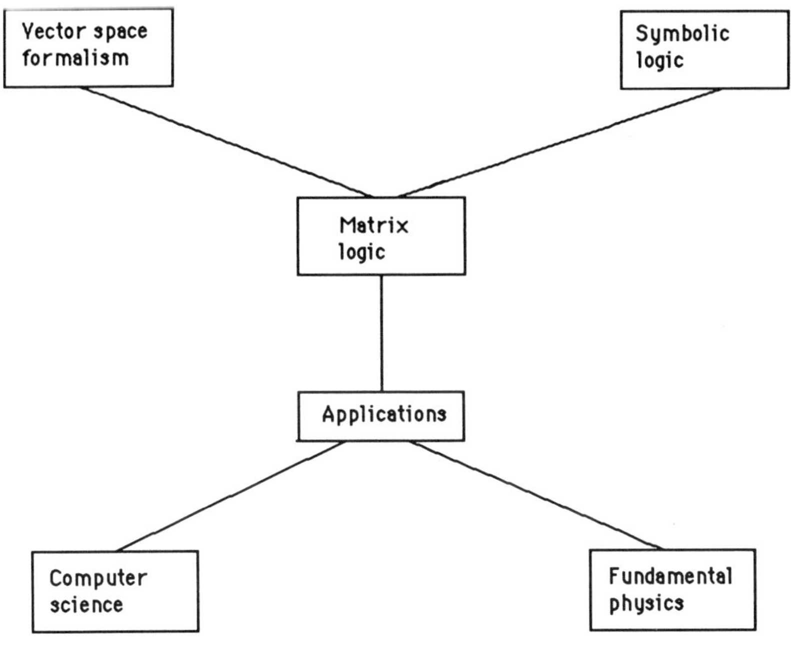

The first chapter surveys a basic knowledge of matrices and matrix algebra, with the purpose of providing elements needed for further reading. The second chapter gives a concise account of the fundamentals of conventional logic, by which we understand all those logic systems in which the logical connectives are not given the status of mathematical objects but are represented by the naive approach that employs truth-tables. Consequently, in the second chapter we treat not only such subjects as first-order-propositional logic and predicate logic, but also multi-valued logic, modal logic, and quantum logic. The second chapter is primarily designed to emphasize the techniques for the transformation and evaluation of logical expressions within the framework of conventional symbolic notation. Stress is laid on the evaluative aspect, whereas general problems of decidability, completeness and others are not treated.

The main import of the book is contained in the third chapter, which provides an elaborate introduction to the matrix formulation of logic. We take a step towards forging a link between formal logic and matrix algebra, and by bridging these two quite different domains, arrive at the matrix formulation of logic. The formalism that is developed displays new deductive and inductive power, since it not only allows one to derive the classical results of conventional logic but also permits their generalization, making available novel techniques of logical inference. The discovery of a link between matrices and logical structures offers the possibility of expressing otherwise poorly defined reasoning processes in terms of numbers in matrix operator equations.

Regarding the third chapter, it is necessary to emphasize that, with the exception of the Dirac brackets and a special treatment of the negation operator, matrix logic does not introduce new symbols for logical connective operators, so that they can be distinguished from ordinary Boolean connectives. This is not accidental. The important fact, which will be substantiated in this book, is that the matrix logic algebra incorporates in itself Boolean logic algebra as a special scalar limit. Therefore, conventional Boolean relations can be formulated, if necessary, in the form of matrix logic equations, with no new symbols for connectives required. The converse, however, is not universally applicable.

Matrix logic does not reject conventional logic. The significance of the new formalism resides in the fact that, in addition to generating new important results, it also contains in itself conventional logic as a special scalar limit. This correspondence principle in logic resembles the reduction of relativity theory to classical mechanics for velocities much smaller than the velocity of light, or is similar to macrophysics obtained as a quantum-mechanical limit when the action quantum converges to zero.

The two concluding chapters of the book focus on several highly important aspects of the application of matrix logic. The fourth chapter outlines a framework for applications of matrix logic in fundamental physics. Since matrix logic permits logical processes to be defined in mathematical form similar or identical to the description of fundamental processes, one may seek a synthesis of logical and physical methods in a unified theory.

A formal theory which adequately expresses the features of the real world must mirror these features in its structure because of the laws of logic. The choice of logic cannot be divorced from real-world constraints and ought to be based on an objective study of propositions themselves, thus ultimately via physical experiment. Bringing logic into the realm of the real-world sciences could make the new formalism a source for novel conceptual ways of classifying elementary fields and particles as well as resolving outstanding problems of fundamental physics.

A reflexive symmetry between matrix logic description and descriptions of the quantum field theory could lead to a unification of major significance whereby it might be possible to achieve the logical description of physical processes and vice versa.

Finally, the fifth chapter of the book discusses applications of matrix logic to computer science, with the main emphasis being placed on new computer architecture and parallel logic processing. With a wide range of computational capabilities, matrix logic promises to have important engineering applications for advanced generation computer systems. The multiplicative properties of matrix logic and the uniformity of its operations make the new formalism a very effective description for multiplicative gates and the cellular type of expandable computer architectures. This is especially suitable for the needs of optical computing on the hardware side, and for "logic crunching" on the software side.

In the fifth chapter we indicate a class of problems in which the multiplicative approach of matrix logic allows a radical optimization, and where the use of matrix processors are expected to be most beneficial for information processing technologies. In addition to providing a foundation for matrix processors, with multiplication as the basic operation, and new group-theoretic insights into parallel computer architectures, matrix logic bids fair to contribute to the field of machine intelligence.

This optimism is based on the fact that the power of direct mathematical computation has been introduced for the first time into the domain of logic valuations. Though we are aware that logic operations form a core of high-level intelligence, our knowledge of high-level algorithms is poor. In our attempts to enhance machine intelligence we are handicapped by the essential fact that the only algorithms we fully understand and are capable of managing are the algorithms of computation, and these alone in their present form are insufficient to handle high-level intelligence functions such as abstraction, induction, memorization, intuition and creativity. But since matrix logic extends the power of computation to the domain of logic valuations, the possibility emerges of expressing high-level intelligence in terms of computation which we already know. With logical connectives being given mathematical form, computational understanding of intelligence finds in matrix logic its genuine closure.

This opens new avenues for developing machine intelligence capabilities as well as for the study of intelligence phenomena and the intelligence code in general.

Finally, the study of intelligence has hitherto been carried out within a logic, specifically by developing and investigating different forms of scalar logic. It has never been considerd that understanding intelligence might require not simply a modification of scalar logic but an entirely new generalization, in the same way as in physics we describe different phenomena with clearly distinct quantities such as scalars, vectors, tensors and others. The important innovation that is introduced and pursued in this book is that we place at the foundation of logic not scalar values but more complex mathematical objects, namely logic vectors and matrix operators, joined eventually into the more general concept of a logic tensor.

In the context of the study of intelligence this result has obliged us to reconsider the role which must be assigned to conventional Boolean logic. Since only the results of logic measurements are expressed in scalar form, Boolean logic, as scalar logic, deals mostly with the ordering of these results in a form suitable for communication. The intricate tensor structure of the reasoning process itself is much more complex and lies largely beyond the reach of conventional logic.

CHAPTER 1

ELEMENTS OF MATRIX CALCULUS

1.1. Inner and outer products

Inner and outer vector products are two fundamental constructs which will play an important role throughout the text.

Consider a bra or row-vector $\langle x|$ whose components are ordered horizontally:

$$\langle x| = (x_1\ x_2\ x_3 \ldots x_n)$$

and a ket or column-vector $|y\rangle$ whose components are ordered vertically:

$$|y\rangle = \begin{pmatrix} y_1 \\ y_2 \\ y_3 \\ \cdot \\ \cdot \\ \cdot \\ y_n \end{pmatrix}$$

The inner product $\langle x \mid y\rangle$ is the following composition of the components of the vectors $\langle x|$ and $|y\rangle$:

$$\langle x \mid y\rangle = (x_1 x_2 x_3 \ldots x_n) \begin{pmatrix} y_1 \\ y_2 \\ y_3 \\ \cdot \\ \cdot \\ \cdot \\ y_n \end{pmatrix} = x_1y_1 + x_2y_2 + x_3y_3 + \ldots x_ny_n = \sum_{i=1}^{n} x_iy_i = \alpha$$

The result of the composition is a number α.

The outer product $|y\rangle\langle x|$ represents the following composition of the components of the vectors $|y\rangle$ and $\langle x|$:

$$|y\rangle\langle x| = \begin{pmatrix} y_1 \\ y_2 \\ y_3 \\ \cdot \\ \cdot \\ \cdot \\ y_n \end{pmatrix} (x_1\ x_2\ x_3 \dots x_n) = \begin{pmatrix} y_1x_1 & y_1x_2 & \cdots & y_1x_n \\ y_2x_1 & y_2x_2 & \cdots & y_2x_2 \\ y_3x_1 & y_3x_2 & \cdots & y_3x_n \\ \cdot & \cdot & \cdots & \cdot \\ \cdot & \cdot & \cdots & \cdot \\ \cdot & \cdot & \cdots & \cdot \\ y_nx_1 & y_nx_2 & \cdots & y_nx_n \end{pmatrix}$$

The result of the composition is an $(n \times n)$ matrix.

Though the result of the first composition and the result of the second are quite different, they are both realized by what is basically an identical operation, that is, the components of the row of the first vector, bra or ket, are multiplied by the corresponding components of the column of the second vector, ket or bra, and if there are several results they are added.
The interaction of two vectors can, depending on the order of operations, generate two different mathematical objects: either numbers or matrices. In this respect, the inner and outer products bridge two important mathematical domains: the domain of numbers and the domain of matrices.
The role of inner and outer products in vector space formalism is indispensable. The outer product serves to define the closure relation which can be employed to construct a vector space. The inner product offers a way of defining the orthogonality of vectors that is both easy and general. Two vectors, neither of which is the null vector, are orthogonal when their inner product is 0:

$$\langle x_i \mid x_j\rangle = 0.$$

We shall make particular use of orthonormal vectors, that is, vectors which are both normalized and orthogonal:

$$\langle x_i \mid x_j\rangle = \delta_{ij}$$

where δ_{ij} is the Kronecker delta defined as follows:

$$\delta_{ij} = \begin{cases} 1 & \text{for} \quad i = j \\ 0 & \text{for} \quad i \neq j \end{cases}$$

In complex vector spaces we employ the more general form of the inner product

$$\langle x \mid y \rangle = \sum_{i=1}^{n} x_i^* y_i \ ,$$

where x_i^* denotes the complex conjugate of x_i.

With some types of vector space, n may tend to infinity and the sum may be replaced by an integral over a continuous parameter.

1.2 Vector space formalism

An associative and commutative set of elements $\{|x_1\rangle, |x_2\rangle, |x_3\rangle, \ldots\}$ constitutes a ket vector space $|V\rangle$ if the following axioms are satisfied:

1. $|x_1\rangle, |x_2\rangle \in |V\rangle \Rightarrow (|x_1\rangle + |x_2\rangle) \in |V\rangle$
2. $|0\rangle \in |V\rangle, |x\rangle + |0\rangle = |x\rangle$
3. $|x\rangle \in |V\rangle \Rightarrow |-x\rangle \in |V\rangle$
4. $|x\rangle \in |V\rangle \Rightarrow c|x\rangle \in |V\rangle$, c-scalar.

A subspace of a vector space $|V\rangle$ is a subset which is itself a vector space. A subset is a subspace if and only if it is closed under addition and scalar multiplication by numbers.

Dual spaces

Given a ket vector space $|V\rangle$ we can define the dual bra vector space $\langle V|$, with a 1–1 correspondence between both spaces. Whatever the vector $|v\rangle$ stands for, the same information is held within $\langle v|$; both are equally informative and neither is to be considered more important or fundamental than the other. The results derived for a ket vector space are automatically applicable to a bra vector space, a fact we shall not mention unless the need arises.

Basis set

The complete vector space can be generated from vectors called basis vectors. A basis for a vector space $|V\rangle$ is an ordered set of vectors

$$\{|x_1\rangle, |x_2\rangle, \ldots, |x_n\rangle\}$$

which span $|V>$ and are linearly independent. The basis vectors may be anything from the unit vectors {i, j, k} of 3-D space to the complete set of eigenvectors of an operator suitable for spanning some Hilbert space. Once the basis is stipulated, we can specify any vector by stating the ordered set of components

$$|x> = a_1 \,|\, e_1> + a_2 \,|\, e_2> + a_3 \,|\, e_3> + \ldots + a_n \,|\, e_n>$$

where $\{a_1, a_2, a_3, \ldots, a_n\}$ is the representation of a vector $|x>$ in the $\{|e_1>, |e_2>, |e_3>, \ldots |e_n>\}$ basis system. The same vector can be described in terms of another basis system $\{|e'_1>, |e'_2>, |e'_3>, \ldots |e'_n>\}$ aligned at some angle to the first:

$$|x> = a'_1 \,|\, e'_1> + a'_2 \,|\, e'_2> + a'_3 \,|\, e'_3> + \ldots a'_n \,|\, e'_n>$$

where $\{a'_1, a'_2, a'_3, \ldots a'_n\}$ is the set of new components providing the representation of the same vector in the new basis system.

As we mentioned, the key property of a basis set of vectors is linear independence. The vectors $|x_1>, |x_2>, \ldots |x_n>$ of a vector space are linearly dependent if there is a relation

$$a_1|x_1> + a_2|x_2> + \ldots + a_n|x_n> = 0 ,$$

where $a_1, a_2, \ldots a_n$ are numbers which are not all zero.

Otherwise the vectors are linearly independent, and any subset of such linearly independent vectors in $|V>$ can be extended to form a basis. Given a set of vectors $\{|x_1>, |x_2>, \ldots |x_n>\}$, we can extend any linearly independent subset to a maximal subset of linearly independent vectors, that is, to a linearly independent subset which is not contained in any other linearly independent subset.

A vector space may not have a finite basis. But if it has a basis with n elements then any other basis also has n elements. The number n characterizes the dimension of the vector space. Any finite set of vectors which span a vector space $|V>$ contains a basis. A maximal subset of linearly independent vectors is automatically a basis. It follows that if $|V>$ has dimension n then any n vectors which span $|V>$ form a basis.

Closure relation

Special importance attaches to representations of vectors in complete orthonormal basis sets, that is the sets of n vectors $\{e_i\}$, $i = 1,n$, in $|V>$ which display orthonormality:

$$<e_i \,|\, e_j> = \delta_{ij} ,$$

and closure:

$$\sum_{i=1}^{n} |e_i\rangle\langle e_i| = I ,$$

where I denotes the unit matrix.
If $\{|e_i\rangle\}$ is a complete orthonormal basis set, then we can express vector $|x\rangle$ in terms of its components as follows:

$$|x\rangle = \sum_{i=1}^{n} x_i\, |\, e_i\rangle$$

where x_i is the orthogonal projection of $|x\rangle$ along the basis vector $|e_i\rangle$:

$$x_i = \langle e_i\, |\, x\rangle .$$

Consequently, the vector $|x\rangle$ can be written

$$|x\rangle = \sum_{i=1}^{n} |e_i\rangle\langle e_i\, |\, x\rangle.$$

By taking a decomposition of $|x\rangle$ relative to some given complete orthonormal basis and summing all the projections of $|x\rangle$ we again get the vector $|x\rangle$, which leads immediately to the closure relation:

$$I = \sum_{i=1}^{n} |e_i\rangle\langle e_i\, | .$$

Thus a closure relation applies if and only if the basis set is complete.

Linear operators

The closure relation which is employed to construct a vector space makes use of the projection operator. Projection operators, like operators in general, can be defined by the ways in which they act upon vectors and later we will frequently encounter situations in which a vector is affected by a matrix operator. We shall here consider linear matrix operators, but operators are not necessarily associated with matrices and are not necessarily linear. The operator A is linear if for all $|x_1\rangle$ and $|x_2\rangle$ from $|V\rangle$:

$$A(c_1 | x_1\rangle + c_2 | x_2\rangle) = c_1 A | x_1\rangle + c_2 A | x_2\rangle ,$$

where c_1, c_2 are numbers.

The operators may cause a rotation, reflection, or displacement of a vector. In generalized vector spaces, the operations may be of a more general nature.

Two operators A and B are said to be equal if for every $|x\rangle \in |V\rangle$:

$$A | x\rangle = B | x\rangle .$$

Similarly, the sum and product of A and B are defined through

$$C = A + B \quad \text{if} \quad C | x\rangle = A | x\rangle + B | x\rangle$$
$$D = AB \quad \text{if} \quad D | x\rangle = A(B | x\rangle) .$$

A linear operator describes a mapping of a vector space into itself. It will be fully defined if any vector $|x\rangle \in |V\rangle$ can be put in correspondence with some other vector $|y\rangle \in |V\rangle$.

This correspondence can be conveniently expressed by the equation

$$|y\rangle = A | x\rangle ,$$

which implies that the operator A associates in a unique fashion with every vector $|x\rangle \in |V\rangle$, a new vector $|y\rangle = A | x\rangle$, $|y\rangle \in |V\rangle$.

Consequently we may present the meaning of A by listing each $|x\rangle$ of $|V\rangle$, and then the corresponding $|y\rangle = A | x\rangle$. But we can much more efficiently define the same operator A by recording what it does to the n basis vectors $\{|e_j\rangle\}$ spanning the vector space $|V\rangle$:

$$A | e_j\rangle = |w_j\rangle .$$

The operator A transforms $|e_j\rangle$ into some other vector $|w_j\rangle$, which can then itself be expanded in the $|e_j\rangle$ basis system. With the aid of the closure relation we obtain

$$A | e_j\rangle = \sum_{i=1}^{n} | e_i\rangle\langle e_i | (A | e_j\rangle) = \sum_{i=1}^{n} |e_i\rangle\langle e_i | w_j\rangle = \sum_{i=1}^{n} |e_i\rangle a_{ij}$$

The inner product of $|w_j\rangle$ with the basis vector $\langle e_i|$ produces a scalar number a_{ij}:

$$<e_i \mid w_j> = <e_i \mid (A \mid e_j>) = a_{ij} .$$

Because no special importance is ascribed either to bra or ket vector space, we must assume that $<e_i \mid A \mid e_j>$ also has an meaning. The operator A was defined through the way it affects the ket basis vectors, delivering a particular set of numbers $\{a_{ij}\}$. But we may equally well consider that the operator A, acting on bra basis vectors $<e_i| A$, produces a unique bra vector whose components are determined by

$$(<e_i \mid A) \mid e_j> = <e_i \mid (A \mid e_j>) = <e_i \mid A \mid e_j> = a_{ij} .$$

Consequently, the priority brackets can be freely deleted and the operator A can operate either to the left or to the right, as the need arises.
Considering our initial equation

$$|y> = A \mid x> ,$$

and employing the closure relation in several places, we may examine the effect of the operator A upon an arbitrary vector $|x>$:

$$\sum_{k}^{n} |e_k><e_k \mid y> = \sum_{kj}^{n} |e_k><e_k \mid A \mid e_j><e_j \mid x>$$

$$\sum_{k}^{n} |e_k>y_k = \sum_{kj}^{n} |e_k>a_{kj}x_j$$

where

$$a_{kj} = <e_k \mid A \mid e_j>$$

At this stage we can project out the components of vector $|y>$ by taking its inner product with the basis vectors $<e_i|$

$$\sum_{k}^{n} <e_i \mid e_k>y_k = \sum_{kj}^{n} <e_i \mid e_k>a_{kj}x_j$$

$$y_i = \sum_{j} a_{ij}x_j ; \qquad i = 1,2,3,...n$$

where the last step is due to the orthonormality of the $\{|e_i>\}$:

$$\langle e_i | e_k \rangle = \delta_{ik}$$

The group of linear transformations

$$y_i = \sum_j a_{ij} x_j$$

is a relationship which tells us how the representation of a vector is affected by the application of an operator. Furthermore, it is one of the sources of the rules of matrix multiplication. If we make column vectors out of the components x_j and y_i, then the n equations of the group of linear transformations can be summarized in matrix notation as follows

$$\begin{pmatrix} a_{11} & \cdots & a_{1n} \\ \cdot & \cdots & \cdot \\ \cdot & \cdots & \cdot \\ \cdot & \cdots & \cdot \\ a_{n1} & \cdots & a_{nn} \end{pmatrix} \begin{pmatrix} x_1 \\ \cdot \\ \cdot \\ \cdot \\ x_n \end{pmatrix} = \begin{pmatrix} y_1 \\ \cdot \\ \cdot \\ \cdot \\ y_n \end{pmatrix}$$

where elements a_{ij} are the n^2 entries into a matrix representation of the operator A in the particular $\{|e_i\rangle\}$ basis system.

It would be a mistake to equate an operator A with a matrix (a_{ij}). When in a fixed basis system we study different operators and relations between them, we can ignore the distinction between an operator and its matrix, since one determines the other. But in another basis system, the same operator A would be represented by another matrix A'. Therefore the set of numbers (a_{ij}) does not characterize the operator A, but a conjunction of the operator and the selected basis system. Only the set of all matrices A, A', A" ... representing an operator in all possible basis systems can be identified with the operator.

1.3 Algebra of matrices

A matrix is an ordered set of numbers:

$$\begin{pmatrix} a_{11} & a_{12} & a_{13} & \cdots & a_{1n} \\ a_{21} & a_{22} & a_{23} & \cdots & a_{2n} \\ \cdot & \cdot & \cdot & \cdots & \cdot \\ a_{m1} & a_{m2} & a_{m3} & \cdots & a_{mn} \end{pmatrix}$$

It is at the same time a simple and fundamental mathematical construct. The theory of matrices, which appeared initially with the aim of dealing, in a concise code, with the coefficients of a system of linear equations, has grown and expanded into an extremely powerful and sophisticated branch of mathematics with numerous applications.

In this chapter we discuss the fundamentals of the formalism of matrix calculus leaving out numerous results but emphasizing those aspects of the formalism which are necessary for a proper understanding of the following chapters.
As we mentioned, the theory of matrices originated in connection with the problem of solving systems of linear equations:

$$\begin{cases} a_{11}x_1 + a_{12}x_2 + \ldots + a_{1n}x_n = y_1 \\ \cdot \quad \cdot \quad \cdot \quad \cdot \quad \cdot \quad \cdot \quad \cdot \quad \cdot \quad \cdot \quad \cdot \quad \cdot \quad \cdot \\ a_{m1}x_1 + a_{m2}x_2 + \ldots + a_{mn}x_n = y_n \end{cases}$$

With the help of matrices the system of equations can be re-written as

$$\begin{pmatrix} a_{11} & a_{12} & \cdots & a_{1n} \\ \cdots & \cdots & \cdots & \cdots \\ a_{m1} & a_{m2} & \cdots & a_{mn} \end{pmatrix} \begin{pmatrix} x_1 \\ \vdots \\ x_n \end{pmatrix} = \begin{pmatrix} y_1 \\ \vdots \\ y_n \end{pmatrix}$$

where the detached coefficients a_{ij} form a matrix A.
This concise notation describes linear transformations from one set of variables to another set. In the theory of vector spaces we encounter similar situations, and that is a reason why the formalism of vector spaces provides an alternative treatment of the theory of matrices.
It soon became evident that the behaviour of the entities A can be studied independently of the problems of linear equations, and can be expressed in a simple and elegant code similar to that of the algebra of ordinary numbers.
Furthermore, matrices not only encode certain relations among the constituent numbers, they also can be viewed as a generalization of ordinary numbers, whereas the algebra of ordinary numbers can be considered as the limiting case of matrices with m = n = 1.
Throughout our discussion it would be of help to be aware of all three aspects of the notion of a matrix: the generalization of numbers, the coefficients of linear equations, and representations of vector space operators.

Equality of matrices

Equality of matrices is defined through equality of their elements. Two matrices A and B are equal if corresponding elements of both matrices are equal:

$$A = B \quad \text{implies} \quad a_{ij} = b_{ij} \,.$$

Addition of matrices

Consider the two vectors |y> and |x>, which are obtained by applying two matrix operators A and B to the same vector |x>:

$$|y\rangle = A\,|\,x\rangle \qquad y_i = \sum_k a_{ik} x_k$$

$$|z\rangle = B\,|\,x\rangle \qquad z_i = \sum_k b_{ik} x_k$$

Vector |y + z> can be computed by applying to |x> matrix operator D, which is obtained by adding matrices A and B:

$$|y + z\rangle = (A + B)\,|\,x\rangle = D\,|\,x\rangle$$

$$y_i + z_i = \sum_k (a_{ik} + b_{ik}) x_k = \sum_k d_{ik} x_k$$

Matrix D = A + B, elements d_{ik} of which are found as sums of the corresponding elements of matrices A and B, is called the sum of the two matrices. Addition of matrices is defined only for matrices of the same order, that is, matrices having an equal number of rows and columns.

Multiplication of matrices

Consider a two-dimensional vector space, and let vector |y> be obtained from some vector |x> by application of an operator B. The components y_1 and y_2 of a vector |y> can be expressed as linear functions of the components x_1 and x_2. We obtain two equations

$$y_1 = b_{11} x_1 + b_{12} x_2$$

$$y_2 = b_{21} x_1 + b_{22} x_2$$

where b_{ij} are some numbers.

The linear operator B is thus defined by the transformation matrix:

$$\begin{pmatrix} b_{11} & b_{12} \\ b_{21} & b_{22} \end{pmatrix}.$$

To derive the rule of the multiplication of matrices consider consecutive applications of operators B and A to a vector |x>.

Let A and B be defined by matrices

$$A = \begin{pmatrix} a_{11} & a_{12} \\ a_{21} & a_{22} \end{pmatrix} \qquad B = \begin{pmatrix} b_{11} & b_{12} \\ b_{21} & b_{22} \end{pmatrix}$$

And let application of operator B to a vector |x> produce a vector |y>, and application of operator A to a vector |y> produce a vector |z>.

We are interested in finding a transformation matrix C which allows direct translation of |x> into |z>:

$$|z\rangle = C\,|\,x\rangle = A \cdot B\,|\,x\rangle.$$

Since

$$y_1 = b_{11}x_1 + b_{12}x_2$$

$$y_2 = b_{21}x_1 + b_{22}x_2$$

and

$$z_1 = a_{11}y_1 + a_{12}y_2$$

$$z_2 = a_{21}y_1 + a_{22}y_2$$

substitution of y_1 and y_2 from the first system of equations into the second will give us

$$z_1 = (a_{11}b_{11} + a_{12}b_{21})x_1 + (a_{11}b_{12} + a_{12}b_{22})x_2$$

$$z_2 = (a_{21}b_{11} + a_{22}b_{21})x_1 + (a_{21}b_{12} + a_{22}b_{22})x_2$$

Hence the transformation matrix

$$C = A \cdot B$$

has the form

$$\begin{pmatrix} c_{11} & c_{12} \\ c_{21} & c_{22} \end{pmatrix} = \begin{pmatrix} a_{11}b_{11} + a_{12}b_{21} & a_{11}b_{12} + a_{12}b_{22} \\ a_{21}b_{11} + a_{22}b_{21} & a_{21}b_{12} + a_{22}b_{22} \end{pmatrix}.$$

We have derived an explicit rule for the multiplication of two matrices:

$$\begin{pmatrix} c_{11} & c_{12} \\ c_{21} & c_{22} \end{pmatrix} = \begin{pmatrix} a_{11} & a_{12} \\ a_{21} & a_{22} \end{pmatrix} \begin{pmatrix} b_{11} & b_{12} \\ b_{21} & b_{22} \end{pmatrix}$$

where

$$c_{ij} = \sum_k a_{ik} b_{kj} .$$

To compute element c_{ij} we multiply corresponding elements of the i-th row of the first matrix A and the j-th column of the second matrix B and add the results obtained.

According to the definition of matrix multiplication, the latter can be performed only if the number of columns k of the first matrix is equal to the number of rows k of the second one: $(m \times k)(k \times n) = (m \times n)$. Multiplications of matrices for which this condition is not satisfied are not defined.

If matrices A and B are rectangular, the only case when the products AB and BA both exist is when A is of the order $(m \times n)$ and B is of the order $(n \times m)$.

Since in the general case

$$AB \neq BA ,$$

where the inequality can concern the matrix elements or the order of the products or both, the ordering of multipliers in matrix algebra is essential and cannot be arbitrarily changed. On the other hand, it is easy to show that associativity and distributivity hold for matrix operations:

$$(A \cdot B)C = A(B \,.\, C)$$

$$A(B + C) = AB + AC$$

The multiplication of matrices can be generalized to incorporate the multiplication of partitioned matrices. In particular the elements of a matrix of the order $(m \times n)$ are submatrices of the order (1×1), the rows are submatrices of the order $(1 \times n)$, and the columns are submatrices of the order $(m \times 1)$.

Since every submatrix in a product matrix is itself the product of two submatrices the general law of the multiplication of partitioned matrices is written

$$C_{ij} = \sum_k A_{ik}B_{kj}$$

where C_{ij}, A_{ik}, B_{kj} are corresponding submatrices.

Where all the submatrices are elements of A and B this rule becomes the ordinary multiplication theorem for matrices.

The addition and multiplication of matrices are more restricted than the addition and multiplication of ordinary numbers. Whereas numbers can always be both added and multiplied, it is quite possible that we can add two matrices but that their multiplication is undefined. For example:

$$\begin{pmatrix} 0 & 1 \\ 1 & 0 \\ 1 & 0 \\ 0 & 1 \end{pmatrix} + \begin{pmatrix} 1 & 0 \\ 0 & 1 \\ 0 & 1 \\ 1 & 0 \end{pmatrix} = \begin{pmatrix} 1 & 1 \\ 1 & 1 \\ 1 & 1 \\ 1 & 1 \end{pmatrix}$$

These matrices can be added but not multiplied. On the other hand the matrices below can be multiplied but not added:

$$\begin{pmatrix} 0 & 1 & 1 & 0 \\ 1 & 0 & 0 & 1 \end{pmatrix} \begin{pmatrix} 0 & 1 \\ 1 & 0 \\ 0 & 1 \\ 1 & 0 \end{pmatrix} = \begin{pmatrix} 1 & 1 \\ 1 & 1 \end{pmatrix}$$

Furthermore the product of the multiplication of two numbers is of the same nature as the multipliers. The multiplication of matrices can deliver a product of a different order from that of the multiplying matrices. This involves rectangular matrices. For example

$$\begin{pmatrix} a_{11} & \cdots\cdots & a_{1n} \\ a_{21} & \cdots\cdots & a_{2n} \end{pmatrix} \begin{pmatrix} b_{11} & b_{12} \\ \vdots & \vdots \\ \vdots & \vdots \\ b_{n1} & b_{n2} \end{pmatrix} = \begin{pmatrix} c_{11} & c_{12} \\ c_{21} & c_{22} \end{pmatrix}$$

Two rectangular matrices $(2 \times n)$ and $(n \times 2)$ produce a square (2×2) matrix.

The rule of matrix multiplication permits the same matrices to be multiplied in the reverse order. But now we get a square $(n \times n)$ matrix:

$$\begin{pmatrix} b_{11} & b_{12} \\ \cdot & \cdot \\ \cdot & \cdot \\ \cdot & \cdot \\ \cdot & \cdot \\ \cdot & \cdot \\ \cdot & \cdot \\ b_{n1} & b_{n2} \end{pmatrix} \begin{pmatrix} a_{11} & \cdots & \cdots & a_{1n} \\ a_{21} & \cdots & \cdots & a_{2n} \end{pmatrix} = \begin{pmatrix} c_{11} & \cdots & \cdots & c_{1n} \\ \cdot & \cdots & \cdots & \cdot \\ \cdot & \cdots & \cdots & \cdot \\ \cdot & \cdots & \cdots & \cdot \\ \cdot & \cdots & \cdots & \cdot \\ \cdot & \cdots & \cdots & \cdot \\ \cdot & \cdots & \cdots & \cdot \\ c_{n1} & \cdots & \cdots & c_{nn} \end{pmatrix}$$

One can easily find a case where two matrices can be multiplied in the direct order, but in the reverse order multiplication is undefined. Thus in spite of being rooted in the domain of numbers, a matrix represents a more general and richer construct.

Mapping between spaces

The property of matrix multiplication which allows one to obtain products of a different order from that of the multipliers can be employed to describe a mapping between spaces of different dimensions. The transformation matrix in such cases can only be rectangular. For example we obtain dimensional contractions:

ket:

$$\begin{pmatrix} 0 & 1 & 0 & 1 \\ 1 & 0 & 0 & 1 \end{pmatrix} \begin{pmatrix} 0 \\ 1 \\ 1 \\ 0 \end{pmatrix} = \begin{pmatrix} 1 \\ 0 \end{pmatrix}$$

bra:

$$\begin{pmatrix} 1 & 0 & 1 & 0 \end{pmatrix} \begin{pmatrix} 0 & 1 \\ 1 & 0 \\ 0 & 0 \\ 1 & 1 \end{pmatrix} = \begin{pmatrix} 0 & 1 \end{pmatrix}$$

The (2×4) matrix operator acting from the left upon the 4-D ket vector contracts it into a 2-D ket vector. The (4×2) matrix operator acting from the right upon the 4-D bra vector contracts it into a 2-D bra vector.

An example of dimensional expansions:

ket:

$$\begin{pmatrix} 1 & 0 \\ 0 & 1 \\ 1 & 0 \\ 0 & 1 \end{pmatrix} \begin{pmatrix} 1 \\ 0 \end{pmatrix} = \begin{pmatrix} 1 \\ 0 \\ 1 \\ 0 \end{pmatrix}$$

bra:

$$(0 \quad 1) \begin{pmatrix} 1 & 0 & 0 & 1 \\ 0 & 1 & 0 & 1 \end{pmatrix} = (0 \quad 1 \quad 0 \quad 1)$$

Dimensional transformations are of certain interest. However, their greatest disadvantage is that we cannot introduce an important class of orthogonal transformations. Orthogonal transformation applies exclusively to square matrices, which provide the mapping of a vector space onto itself.

Square matrices

A matrix of the order $(n \times n)$ with an equal number of rows and columns is called a square matrix. These are of particular interest for our discussion, and especially matrices of the order (2×2). Matrices of order (2×2) represent not only the most basic case of square matrices but also a case of special importance for the formalism developed in the following chapters. Assuming that the operation of multiplication is defined, (2×2) square matrices can be derived as matrix products of corresponding rectangular matrices of the order $(2 \times n)$ and $(n \times 2)$, or as outer products of bra and ket 2-D vectors, for example:

$$\begin{pmatrix} 1 \\ 0 \end{pmatrix} (1 \quad 0) = \begin{pmatrix} 1 & 0 \\ 0 & 0 \end{pmatrix} .$$

A major advantage of square matrices is that these can be both added and multiplied, thus providing a basis for developing an algebra of matrices in many respects reminiscent of the algebra of ordinary numbers, however with some important differences.

Commutativity. The result of the multiplication of numbers does not depend on the order in which the multiplication is performed. When

$$ab = ba$$

we say that commutativity of multiplication holds. With matrix multiplication this is not generally the case. Some matrices commute, some do not.

Matrix diagonals

With respect to square matrices, we define the principal diagonal, propagating from the upper left corner of the matrix to the lower right corner; and the secondary diagonal, propagating from the lower left corner to the right upper corner:

$$\begin{pmatrix} a_{11} & \cdots & a_{n1} \\ \vdots & \times & \vdots \\ a_{n1} & \cdots & a_{nn} \end{pmatrix}$$

The principal diagonal is formed by the set of elements $\{a_{ii}\}$; the secondary diagonal is formed by the set of elements $\{a_{i,n-i+1}\}$.

Though the above explanatory picture appears always to have one common element, the principal and secondary diagonals may consist of two separate subsets, $\{a_{ii}\} \cap \{a_{i,n-i+1}\} = \emptyset$ if the order of the matrix is even, or they have one common element if the order of the matrix is odd.

The notion of diagonals is involved in the definitions of important types of matrices and matrix operations.

Symmetric matrix

A square matrix is defined as symmetric if the elements located symmetrically to each other with respect to the principal diagonal are equal:

$$a_{ij} = a_{ji}$$

Thus the matrices

$$\begin{pmatrix} 0 & 1 \\ 1 & 1 \end{pmatrix} \text{ and } \begin{pmatrix} 1 & 0 \\ 0 & -1 \end{pmatrix}$$

are examples of symmetric matrices, whereas the matrices

$$\begin{pmatrix} 1 & 1 \\ 0 & 1 \end{pmatrix} \text{ and } \begin{pmatrix} 0 & -1 \\ 1 & 0 \end{pmatrix}$$

are non-symmetric ones.

Anti-symmetric matrix

An anti-symmetric matrix occurs when the elements located symmetrically with respect to the principal diagonal are equal but have the opposite sign

$$a_{ij} = -a_{ji}.$$

This implies that all the elements of the principal diagonal in the anti-symmetric matrix are equal to zero:

$$a_{ii} = 0.$$

An example of an anti-symmetric matrix is given below

$$A = \begin{pmatrix} a_1 & a_2 & a_3 & a_4 \\ -a_2 & a_1 & -a_4 & a_3 \\ -a_3 & a_4 & a_1 & -a_2 \\ -a_4 & -a_3 & a_2 & a_1 \end{pmatrix}, \qquad a_1 = 0$$

Matrices whose elements are symmetric across both diagonals we define as bisymmetric matrices. Consider these examples of bisymmetric matrices:

$$\begin{pmatrix} 1 & 0 \\ 0 & 1 \end{pmatrix} \quad \begin{pmatrix} 0 & 1 \\ 1 & 0 \end{pmatrix} \quad \begin{pmatrix} 0 & 0 \\ 0 & 0 \end{pmatrix} \quad \begin{pmatrix} 1 & 1 \\ 1 & 1 \end{pmatrix}$$

Diagonal matrix

A matrix is defined as a diagonal matrix if all its elements are zero, except for those from the main diagonal:

$$a_{ij} = 0 \qquad \text{for} \quad i \neq j$$
$$a_{ij} \neq 0 \qquad \text{for} \quad i = j$$

For example

$$\begin{pmatrix} 1 & 0 & 0 & 0 \\ 0 & 1 & 0 & 0 \\ 0 & 0 & 1 & 0 \\ 0 & 0 & 0 & 1 \end{pmatrix}$$

is a diagonal matrix. Clearly, the above diagonal matrix is both a symmetric and a bisymmetric matrix. A fundamental property of diagonal matrices is that the product of diagonal matrices A and B complies with the commutativity rule AB = BA, and again yields a diagonal matrix.

Of special importance is a diagonal matrix whose diagonal elements are the same number α, a scalar matrix α.

Multiplication by a scalar

Multiplication of a matrix A by a scalar α is obtained by multiplying each element of the matrix by α. The rule for multiplication of a matrix by a scalar can be defined in two equivalent ways: through the addition or through the multiplication of matrices. Indeed, multiplication of a matrix by a scalar α is equal to its product with a scalar matrix α:

$$\alpha \begin{pmatrix} a_{11} & a_{12} \\ a_{21} & a_{22} \end{pmatrix} = \begin{pmatrix} \alpha & 0 \\ 0 & \alpha \end{pmatrix} \begin{pmatrix} a_{11} & a_{12} \\ a_{21} & a_{22} \end{pmatrix} = \begin{pmatrix} \alpha a_{11} & \alpha a_{12} \\ \alpha a_{21} & \alpha a_{22} \end{pmatrix}.$$

From the rule of the addition of matrices it follows that multiplication by a scalar α is equivalent to the addition of α identical matrices:

$$\alpha \begin{pmatrix} a_{11} & a_{12} \\ a_{21} & a_{22} \end{pmatrix} = \begin{pmatrix} \alpha a_{11} & \alpha a_{12} \\ \alpha a_{21} & \alpha a_{22} \end{pmatrix}.$$

The number α can take the form of the operators of differentiation or integration. In such a case, multiplication by α is equivalent to the differentiation or integration of a matrix.

Multiplication of a matrix A by the scalar −1 produces matrix −A, all elements of which have signs opposite to those of the initial matrix A.
In this respect the subtraction of matrices can be defined via addition with prior multiplication of the subtrahend by −1:

$$A - B = A + (-1)B = A + (-B).$$

Unit matrix and null matrix

The ordinary scalars 1 and 0 find their matrix correlants in the form of a unit matrix and a null matrix respectively. A unit matrix I is a diagonal symmetric matrix where the elements on the main diagonal are equal to 1. This can be expressed as

$$a_{ij} = \delta_{ij}$$

where δ_{ij} is the Kronecker index.
A unit matrix, an example of which we have already given above, plays the same role in matrix algebra as the number 1 in ordinary algebra. A multiplication by a unit matrix is an identity operation for an arbitrary matrix A

$$IA = AI = A\ ,$$

and an arbitrary vector $|x\rangle$ or $\langle x|$,

$$I\,|\,x\rangle = |x\rangle$$

Indeed,

$$x_i = \sum_j \delta_{ij} x_j$$

A null matrix $\hat{0}$ is a matrix all of whose elements are equal to null; for example, the (4×4) null matrix:

$$\hat{0} = \begin{pmatrix} 0 & 0 & 0 & 0 \\ 0 & 0 & 0 & 0 \\ 0 & 0 & 0 & 0 \\ 0 & 0 & 0 & 0 \end{pmatrix}.$$

In matrix calculus the null matrix $\hat{0}$ plays a role similar to that of the number 0 in ordinary algebra. $\hat{0}\,A = A\,\hat{0} = \hat{0}$. However, contrary to ordinary algebra, where ab = 0 implies strictly that either a = 0 or b = 0, the product of two matrices can be a null matrix even if both of them differ from a null matrix. Consider these examples:

$$\begin{pmatrix} 1 & 1 \\ -1 & -1 \end{pmatrix}\begin{pmatrix} -1 & 1 \\ 1 & -1 \end{pmatrix} = \begin{pmatrix} 0 & 0 \\ 0 & 0 \end{pmatrix}$$

$$\begin{pmatrix} 1 & 0 \\ 0 & 0 \end{pmatrix}\begin{pmatrix} 0 & 0 \\ 0 & 1 \end{pmatrix} = \begin{pmatrix} 0 & 0 \\ 0 & 0 \end{pmatrix}$$

$$\begin{pmatrix} 0 & 0 \\ 1 & 0 \end{pmatrix}\begin{pmatrix} 0 & 0 \\ 1 & 0 \end{pmatrix} = \begin{pmatrix} 0 & 0 \\ 0 & 0 \end{pmatrix}$$

An important difference between a unit matrix and a null matrix is that a unit matrix is defined only for square matrices whereas a null matrix can be of any order, including rectangular matrices.

The powers of a matrix

In pointing out that matrix multiplication is not universally commutative we come upon an important distinction between matrix algebra and ordinary scalar algebra. However, associativity of multiplication is true in both algebras. A(BC) is identical with (AB)C, both yielding, though by a different order of operations, the same product matrix.

Because a square matrix can be multiplied by itself, by virtue of the associativity of matrix multiplication, the powers $A^2, A^3, \dots, A^n$ can be defined without ambiguity.
Raising a matrix to the power 2 can produce the whole spectrum of results:

$$A^2 = F$$

$$B^2 = B$$

$$C^2 = O$$

$$D^2 = I$$

$$E^2 = E^{-1}$$

where F is some new matrix, $B^2 = B$ represents an important class of idempotent matrices, C is an annihilating matrix, D is characterized by being the inverse of itself, and E^{-1} is an inverse matrix.

Raising a matrix to a negative power is a very important operation. Because division by a matrix is defined via multiplication by an inverse matrix, the introduction of negative powers allows one to complete matrix algebra since it adds to the addition and multiplication of matrices the operation of division.

Inverse matrix

Much as with ordinary numbers, the multiplication of a matrix A by its inverse A^{-1} produces a unit matrix:

$$AA^{-1} = A^{-1}A = I .$$

If matrix operator A defines a transformation of a vector $|x\rangle$ into a vector $|y\rangle$

$$|y\rangle = A\,|\,x\rangle ,$$

then the inverse of A is some matrix operator A^{-1} which cancels the effect of A upon the vector $|x\rangle$:

$$A^{-1}\,|\,y\rangle = A^{-1}A\,|\,x\rangle = I\,|\,x\rangle = |x\rangle .$$

Thus the matrix inverse A^{-1} defines an inverse transformation

$$|x\rangle = A^{-1}\,|\,y\rangle .$$

The matrix elements of A^{-1} are computed by the formula

$$a_{ij}^{-1} = \frac{(-1)^{i+j} A_{ij}}{\det A}$$

where det A is the determinant of matrix A and A_{ij} the minor determinant multiplied by the factor $(-1)^{i+j}$.

It is evident that the computing of the matrix inverse assumes that

$$\det A \neq 0 .$$

Otherwise inverse A^{-1} does not exist and matrix A is called a singular matrix. An inverse matrix is analogous to an inverse number. The analogy is, however, incomplete. Each number $a \neq 0$ has an inverse, whereas there exist non-null matrices - singular matrices - which do not have inverses. A singular matrix is a matrix with a determinant equal to zero. For example

$$\begin{vmatrix} 1 & 1 \\ 1 & 1 \end{vmatrix} = 0 , \quad \begin{vmatrix} 0 & 0 \\ 1 & 1 \end{vmatrix} = 0 , \quad \begin{vmatrix} 0 & 0 \\ 1 & 0 \end{vmatrix} = 0 .$$

Straight brackets are used to denote the determinant, while the matrix is denoted by curve brackets.

The most simple procedure for computing determinants is that for (2×2) matrices; this is found by cross-multiplication:

$$\begin{vmatrix} a & b \\ c & d \end{vmatrix} = ad - cb .$$

Consider determinants and inverses for the following matrices:

$$\begin{vmatrix} 1 & 1 \\ 0 & 1 \end{vmatrix} = 1 , \quad \begin{pmatrix} 1 & 1 \\ 0 & 1 \end{pmatrix}^{-1} = \begin{pmatrix} 1 & -1 \\ 0 & 1 \end{pmatrix}$$

$$\begin{vmatrix} 1 & 0 \\ 1 & 1 \end{vmatrix} = 1 , \quad \begin{pmatrix} 1 & 0 \\ 1 & 1 \end{pmatrix}^{-1} = \begin{pmatrix} 1 & 0 \\ -1 & 1 \end{pmatrix} .$$

The general determinant is the extension of such cross-multiplication from a (2×2) matrix to an $(n \times n)$ matrix.

The inverse of a symmetric matrix is a symmetric matrix; the inverse of a diagonal matrix is a diagonal matrix. This simplifies the computation of inverses for these matrices. For example the inverse of a diagonal matrix

$$A = \begin{pmatrix} a & & & 0 \\ & b & & \\ & & c & \\ 0 & & & d \end{pmatrix}$$

is found immediately as

$$A^{-1} = \begin{pmatrix} \frac{1}{a} & & & \\ & \frac{1}{b} & & 0 \\ & & \frac{1}{c} & \\ 0 & & & \frac{1}{d} \end{pmatrix}.$$

The equality of the determinants of two matrices does not imply the equality of those matrices; the latter could even be of a different order.

Finally, it might be useful to point out that the inverse of the multiplication of matrices is the multiplication of inverses taken in reverse order:

$$(ABC \ldots N)^{-1} = N^{-1} \ldots C^{-1}B^{-1}A^{-1} .$$

Transpose matrix

Computation of the matrix inverse involves the recombination of the rows and columns of a matrix. This important operation is called matrix transposition and a matrix obtained in such a way is called the transpose matrix.

The transpose of a matrix A is a matrix $\widetilde{A}$, obtained from A by the recombination of rows and columns against the principal diagonal

$$\widetilde{a}_{ij} = a_{ij} .$$

The two inverses given above are matrix transposes of each other:

$$\widetilde{\begin{pmatrix} 1 & -1 \\ 0 & 1 \end{pmatrix}} = \begin{pmatrix} 1 & 0 \\ -1 & 1 \end{pmatrix}.$$

Clearly, the transposition of a transposition is an identity operation which keeps the matrix unchanged:

$$\widetilde{\widetilde{a}}_{ij} = a_{ij} .$$

However, there are classes of square matrices which remain unchanged under a single transposition:

$$\tilde{a}_{ij} = a_{ij} .$$

This is a characteristic property of symmetric matrices. For example:

$$\widetilde{\begin{pmatrix} 1 & 0 \\ 0 & 0 \end{pmatrix}} = \begin{pmatrix} 1 & 0 \\ 0 & 0 \end{pmatrix} ; \quad \widetilde{\begin{pmatrix} 1 & 1 \\ 1 & 0 \end{pmatrix}} = \begin{pmatrix} 1 & 1 \\ 1 & 0 \end{pmatrix} .$$

The transpose of the multiplication of two matrices $\tilde{C} = \widetilde{AB}$ is the multiplication of the transposes in reversed order

$$\tilde{C} = \widetilde{AB} = \tilde{B}\tilde{A} .$$

Matrix $\widetilde{AB}$ is obtained by multiplying rows of A on columns of B, and the recombination of rows and columns in the resulting matrix C. But the same outcome emerges from the multiplication of the columns of B, i.e. rows of $\tilde{B}$ with rows of A, i.e. columns of $\tilde{A}$.
This rule is easily generalized for an arbitrary number of matrices:

$$\widetilde{A\,B\,C \ldots N} = \tilde{N} \ldots \tilde{C}\tilde{B}\tilde{A} .$$

Converse transposition

For reasons which will surface in the following chapters we shall introduce an operation of converse transposition or reflection. The reflection of a matrix will be denoted by $\overline{A}$ and defined in a way similar to transposition. The reflection operation also involves the recombination of rows and columns, but now not against the principal diagonal but against the secondary diagonal of the matrix. We have for a matrix transpose

$$\tilde{A} = \widetilde{\begin{pmatrix} a & b \\ c & d \end{pmatrix}} = \begin{pmatrix} a & c \\ b & d \end{pmatrix}$$

whereas for reflection

$$\overline{A} = \overline{\begin{pmatrix} a & b \\ c & d \end{pmatrix}} \begin{pmatrix} d & b \\ c & a \end{pmatrix}$$

The rule of the reflection of the multiplication of matrices is similar to the rule of matrix transposition:

$$\overline{A\,B\,C\,\ldots\,N} = \overline{N}\ldots\overline{C}\;\overline{B}\;\overline{A}\,.$$

For (2×2) matrices, joint execution of reflection and transposition is equal to joint execution of row transposition and column transposition.

$$\tilde{\overline{A}} = \overset{RC}{A}\,,$$

where

$$\overset{R}{A} = \overset{R}{\begin{pmatrix} a & b \\ c & d \end{pmatrix}} = \begin{pmatrix} c & d \\ a & b \end{pmatrix},$$

$$\overset{C}{A} = \overset{C}{\begin{pmatrix} a & b \\ c & d \end{pmatrix}} = \begin{pmatrix} b & a \\ d & c \end{pmatrix}.$$

We also can define more subtle crossing-over transpositions:

<u>ket</u>: $$\hat{A} = \widehat{\begin{pmatrix} a & b \\ c & d \end{pmatrix}} = \begin{pmatrix} b & c \\ d & a \end{pmatrix},$$

<u>bra</u>: $$\check{A} = \overset{\vee}{\begin{pmatrix} a & b \\ c & d \end{pmatrix}} = \begin{pmatrix} d & c \\ a & b \end{pmatrix}$$

with $\hat{\hat{\hat{\hat{A}}}} = A$ and $\check{\check{\check{\check{A}}}} = A$.

Crossing-over transpositions are partially row and partially column transpositions, which are expressed as the square of the crossing-over operations:

$$\overset{\mathrm{R}}{\mathrm{A}} = \hat{\hat{\mathrm{A}}} \quad \text{and} \quad \overset{\mathrm{C}}{\mathrm{A}} = \check{\check{\mathrm{A}}} .$$

A new operation of converse transposition allows us to introduce an important diagonal principle, which will re-appear in various forms throughout the rest of this book and which establishes a relation between transposition and converse transposition:

$$\overset{\sqcap}{\mathrm{A}} = \rceil \tilde{\mathrm{A}} \rceil$$

$$\tilde{\mathrm{A}} = \rceil \overset{\sqcap}{\mathrm{A}} \rceil$$

where $\rceil$ denotes the matrix complement of the unit matrix:

$$\rceil = \bar{\mathrm{I}} = \overset{\mathrm{R}}{\mathrm{I}} = \overset{\mathrm{C}}{\mathrm{I}} .$$

Since $\rceil$ is an orthogonal matrix, transposition and reflection of a (2×2) matrix are related via the orthogonal transformation.

1.4 Orthogonal transformations

Among the different transformations of vector spaces induced by matrix operators, special interest attaches to those transformations that leave the norm of the vectors intact, the orthogonal transformations. In a vector space an orthogonal transformation changes one orthonormal basis into another orthonormal basis. Conversely, if a transformation changes an orthonormal basis into another orthonormal basis, it is an orthogonal transformation.

For a transformation to be orthogonal it is sufficient that the norm of a vector is conserved, that is, it is sufficient that the norm of a vector is invariant under such a transformation.

Consider a vector $|p\rangle$, with the square of the norm expressed by the inner product

$$\|p\|^2 = \langle p \mid p \rangle .$$

In a new space, obtained by orthogonal transformation θ, the square of the norm of this vector is expressed as

$$\|q\|^2 = \langle q \mid q \rangle .$$

Then

$$\langle p \mid p \rangle = \langle q \mid q \rangle .$$

Since

$$|p\rangle = \theta \mid q\rangle$$

and

$$\langle p| = \langle q \mid \tilde{\theta} ,$$

we have

$$\langle p \mid p \rangle = \langle q \mid \tilde{\theta}\, \theta \mid q \rangle = \langle q \mid q \rangle .$$

Thus

$$\tilde{\theta}\, \theta = I$$

and

$$\tilde{\theta} = \theta^{-1} .$$

Consequently, for the matrix of an orthogonal transformation, the transpose is equal to the inverse. If the columns and rows of a square matrix form a set of orthonormal vectors, then a matrix is orthogonal.

We now consider two vectors $|x_1\rangle$ and $|x_2\rangle$ which, in some orthonormal basis, are related by the matrix operator A:

$$|x_2\rangle = A \mid x_1\rangle.$$

Performing an orthogonal transformation θ of the given basis, we obtain a new orthonormal basis in which vector $|x_1\rangle$ and $|x_2\rangle$ are expressed as $|y_1\rangle$ and $|y_2\rangle$ respectively.

The task is to define a matrix operator B, which relates vectors $|y_1\rangle$ and $|y_2\rangle$ in the new basis, obtained by orthogonal transformation:

$$|y_2> = B \mid y_1>.$$

Since

$$|x_2> = A \mid x_1>$$
$$|x_1> = \theta \mid y_1>$$
$$|x_2> = \theta \mid y_2>$$

we can write

$$\theta \mid y_2> = A\theta \mid y_1>$$

Multiplying from the left both sides of the matrix equation by θ^{-1},we obtain

$$|y_2> = \theta^{-1} A\theta \mid y_1> .$$

From this it follows that

$$B = \theta^{-1} A\theta .$$

If transformation θ is orthogonal, then

$$\theta^{-1} = \tilde{\theta} .$$

Consequently, the orthogonal transformation of matrix operator A is expressed as

$$B = \tilde{\theta}\, A\, \theta.$$

The operations of the orthogonal transformation can be generalized for complex vector spaces. The counterpart of an orthogonal matrix in a complex vector space is the unitary matrix U, satisfying the condition:

$$U^+ U = U U^+ = I ,$$

where U^+ is the conjugate transpose of U,

$$U^+ = U^{-1} .$$

The inverse of the unitary operator is identical with its conjugate transpose. In the complex vector space unitary transformations thus take the form:

$$B = U^{+} A U .$$

CHAPTER 2

FUNDAMENTALS OF CONVENTIONAL LOGIC

In the most general sense logic can be interpreted as a fundamental code of reasoning that embodies intelligent processes. Symbolic or mathematical logics represent various attempts to formalize different aspects of logic, and in this way to reduce all human reasoning to unambiguous mathematical notation.

The birth of mathematical logic is associated with the work of George Boole, who formulated its initial principles as propositional symbolic logic and created the branch of mathematics known today as Boolean algebra. The development of mathematical logic became possible upon realization that if the simple propositions of logic could be represented by precise symbols, then the relation between them could be read as precisely as algebraic equations.

At present symbolic logic is a vast field, in which different formalizations enjoy different degrees of success. Seeking initially to provide a formal embodiment of intelligent processes, symbolic logic has found important theoretical and practical applications beyond its initial classical domain that of the investigation of reasoning.
The study of logic as well as the study of mathematical logic is far from completion. Mathematical logic continues to be developed, to a great extent under the influence of problems arising in applications.

Propositional Boolean logic restricts itself to the study of true and false logic expressions, with true and false representing the so-called binary alphabet of truth-values. A logic designed over such an alphabet is called a two-valued or binary logic. This convenient simplicity of the alphabet of truth-values, however, imposes severe limitations on the objects of logical inquiry, since it disregards all those expressions which cannot be defined in terms of true and false, such as those which are undecided, undetermined, unknown, meaningless etc.
Investigation of these and other intelligent properties of logical discourse is attempted in more general forms of conventional logic: multivalued logics, modal logics. In this chapter we shall first consider first-order logic and then overview briefly some properties of higher-order logics.

2.1 First-order logic

Atomic and compound expressions

A distinction is made between elementary and compound logic expressions. Elementary expressions are also called atomic expressions and are denoted by small letters: x, y, z, . . ., the compound expressions being denoted by capitals: X, Y, Z, ... The compound expressions are formed from atomic expressions, and just like atomic expressions, they can also be identified as either true or false.
To form a compound expression in symbolic logic we introduce symbols for connectives. These "connect" elementary atomic expressions, holding them in a linear string. Traditionally, connectives are defined by so-called truth-tables, which specify the truth-values of the compound depending on the truth-values of the constituent atomic expressions. The notion of connectives lies at the very foundation of logic.
Together with atomic expressions, axioms and rules of inference they actually represent those fundamentals from which a logic calculus can be designed.
Conventional logic distinguishes between unary and binary connectives. As the terminology indicates, the unary connective acts upon a single logic expression, either atomic or compound; the binary connective acts between two logic expressions, either atomic or compound.

Unary connectives

Affirmation or identity

It seems natural to begin the description of logic connectives with the identity connective, denoted by I. Affirmation of x is true if x is true and false if x is false. Thus the truth-table for the unary connective identity Ix is

x	Ix
f	f
t	t

where "t" both here and henceforth stands for "true", "f" standing for "false".

Negation

The unary connective negation, denoted by $\rceil$, represents a fundamental logic operation. If x is an expression, $\rceil$x (or $\bar{x}$) symbolises the negation of x and is read "not x". The truth-table of negation is thus

x	$\rceil$x
f	t
t	f

The negation of x is false if x is true , and true if x is false. Clearly, negation is the opposite of identity.

Logic constants

Two more unary connectives are defined in binary logic. These are the logic constants False and True, with the following truth-tables:

x	Fx
f	f
t	f

x	Tx
f	t
t	t

Independently of the truth-value of x, the logic constant in one case always obtains the truth-value f, in the other case the truth-value t.

Binary connectives

Two atomic expressions x and y may generate $2^2 = 4$ combinations of truth-values; ff, tf, ft, tt. Each combination set can yield $2^{2^2} = 16$ different truth-tables which also consist of the truth-values either true or false. With each of these 16 truth-tables conventional logic associates a corresponding binary Boolean connective.

Though, according to Jean Piaget, our intelligence makes use of all 16 logical operations, in most literature on logic only a few binary connectives are selected as theoretically and practically relevant: conjunction, disjunction, implication and equivalence. We now consider the truth-tables for this standard set of connectives.

Conjunction $\wedge$

The properties of conjunction appear to mirror some properties of the word "and": an expression of the form $x \wedge y$ is read as "x and y". The conjunction $x \wedge y$ is true if x and y are both true and is otherwise false. The truth-table for conjunction has the form

x	y	$x \wedge y$
f	f	f
f	t	f
t	f	f
t	t	t

Conjunction is a symmetric connective and the order of appearance of the atomic expressions is irrelevant for the result:

$$x \wedge y = y \wedge x \, .$$

We say x and y commute under conjunction.

Disjunction $\vee$

The disjunction of x and y is false if x and y are both false and is otherwise true. In ordinary language the connective $\vee$ corresponds to the word "or" which we use to separate two expressions. However, it is necessary to recognize that in language the disjunction "or" is applied in two different ways: it may mean that only x or only y, but not both, is true - the exclusive usage; it may also mean that either one of them or both are true - the inclusive usage. Sometimes exclusive disjunction is called strong alternation: $\dot{\vee}$, and inclusive disjunction, weak alternation: $\vee$. Disjunction is here understood in a weak sense and thus will be defined by the following truth-table:

x	y	$x \vee y$
f	f	f
f	t	t
t	f	t
t	t	t

It so happens that the use conventional logic makes of exclusive disjunction $\dot{\vee}$ is negligible. Even a symbol answering to it is seldom introduced. To assert the exclusive disjunction of two expressions x and y, conventional logic goes via substitutes like

$$x \dot{\vee} y = (x \wedge \bar{y}) \vee (\bar{x} \wedge y) .$$

Strong alternation is considered to be not as basic to our formal languages as weak alternation. In the following chapter we shall take issue with this viewpoint and defend quite the opposite thesis, namely that strong alternation, being re-defined in a special way, may well be the most fundamental connective of all.

Finally, note that disjunction, like conjunction, is a symmetric connective:

$$x \vee y = y \vee x .$$

We say x and y commute under disjunction.
It is necessary to emphasize that commutation of disjunction and conjunction holds only when we are concerned with the truth status of x and y and not with their meanings. Commutation, however, can fail to apply to causally dependent logical events occurring in time. In such cases commuting conjunction and disjunction must be replaced by sequential conjunction and sequential disjunction, respectively, where x and y are not permutable.

Implication $\rightarrow$

Implication, "if x then y" or "x implies y", is denoted by $x \rightarrow y$, where x is called the antecedent and y is called the consequent. By definition, implication is always true if the consequent is true or if the antecedent is false. This is quite different from the causal relation and one should not interpret logic implication in terms of cause and effect in a physical sense. Though logic implication resembles the implication of ordinary discourse, it is nevertheless an imperfect approximation to the latter. But it does preserve the most important feature of ordinary reasoning: a true implication completely rules out the possibility of a true antecedent implying a false consequent. The truth-table of implication is thus defined as follows:

x	y	$x \rightarrow y$
f	f	t
f	t	t
t	f	f
t	t	t

The truth-table for implication is a common source of confusion, which arises from not clearly separating the causal and logical aspects of implication. The difficulty arises when a false antecedent implies a true consequent. Since this can lead to meaningless statements, some people tend to think that such expressions are not useful. This is not so. Whereas the implication we are familiar with in common usage is based upon some connection in meaning between the antecedent and the consequent, a logic implication can represent only a relationship between the truth status of the antecedent and that of the consequent and not a relation between their meanings. In this context the significance of implication is that it enables the truth of y to be inferred from the truth of x, and nothing in particular to be inferred from the falsity of x.

Equivalence ↔

In common language the analogue of the logic connective equivalence is represented by "if and only if". In logic, equivalence $x \leftrightarrow y$ is true if and only if x and y have identical truth-values:

x	y	$x \leftrightarrow y$
f	f	t
f	t	f
t	f	f
t	t	t

The truth-table for equivalence presented above can be derived from the truth-tables of conjunction and implication since

$$(x \leftrightarrow y) = (x \rightarrow y) \wedge (y \rightarrow x) ,$$

or from the truth-tables for conjunction, disjunction and negation since

$$(x \leftrightarrow y) = (x \wedge y) \vee (\bar{x} \wedge \bar{y}) .$$

The above truth-tables for logic connectives represent an important operational base of symbolic logic.

Without these simple, yet elegant constructs it would be quite difficult to imagine the successful development of mathematical logic we observe today.

Truth-tables of compound expressions

In the preceding sections we considered the truth-tables for simple logic expressions containing not more than two constituent atomic expressions. In an analogous way we can establish the truth-tables for compound expressions which consist of an arbitrary number of atomic expressions and connectives.

Each of the compound expressions can also be characterized by its corresponding truth-table. The truth-table of a compound containing n atomic expressions has 2^n rows, whereas the number of different possible truth-tables becomes 2^{2^n}, and rapidly increases if n increases. The following is the truth-table for the logic expression

$$(x \wedge y) \rightarrow (z \wedge x) ,$$

containing three different atomic expressions:

x	y	z	$(x \wedge y) \rightarrow (z \vee x)$
f	f	f	t
f	f	t	t
f	t	f	t
f	t	t	t
t	f	f	t
t	f	t	t
t	t	f	f
t	t	t	t

Logic functions

Since the truth-value of a logic expression is fully determined by the truth-value of its constituent atomic expressions, every logic expression can be viewed as a certain Boolean function, the arguments of which, independently of each other, obtain the values "true" or "false". The same applies to the function itself. Because the arguments of logic functions and functions themselves may take only two different values, there are a finite number of different argument sets for a function, and a logic function is uniquely defined by a table listing all the possible argument sets and corresponding function values.

Since, in symbolic logic, we are not interested in the content of an expression as such but only in its truth status, we can characterize a logic expression by assigning to it some numerical value. The usual way is to assign to true expressions the numerical value 1, and to false expressions the numerical value 0.

After the given assignments each atomic expression and each compound expression emerge in new roles, namely in the roles of a logic variable and a logic function respectively, both defined on the binary set

$$E_2 = \{0, 1\} .$$

On this line of reasoning any logic function $L(x_1, x_2, \ldots x_n)$, where x_i, $i = 1,n$, are logic variables, represents a mapping

$$\underbrace{E_2 \times E_2 \times \ldots \times E_2}_{n \text{ times}} \rightarrow E_2 ,$$

with both the logic variables and logic function itself defined on the binary set E_2:

$$x_i \in E_2$$
$$L(x_1, x_2, \ldots x_n) \in E_2 .$$

The number of possible Boolean functions from n variables rapidly increases if n increases. For n = 1 there are 4 different functions, for n = 2, 16 functions, for n = 3 we can construct 256 functions, and for n = 4, 65536 functions.

If one wanted to find out whether some functions have some property, it would seem natural to try to scan the given set of functions, which we know in advance is finite. However, even for relatively small values of n, such a search becomes quite difficult even with the support of a computer. The truth-table of a Boolean function for n = 10 contains 1024 rows, and for n = 20 becomes practically unobservable.

Expressing compound Boolean functions via more simple functions, in other words, considering a function of a function of the form

$$L(L_1(L_2(L_3\ldots(x,y)))) \ldots) ,$$

we can end up with functions which are considered as elementary Boolean functions, similar to sin x or x^2 considered as elementary in analysis. In the first section of this chapter we already gave a semiformal description of some of these functions. Now we shall present the complete set of elementary Boolean functions and underline some of their properties relevant to our discussion.

Boolean functions of one argument

There are four Boolean functions of one argument: identity I(x), negation ⌉(x), and the logic constants 1(x) and 0(x). The chart below shows that the structure of these functions is identical with the structure of the truth-tables for the unary connectives we considered above, except that the truth-values t and f are now replaced by the numerical values 1 and 0 respectively:

x \ L(x)	I(x)	⌉(x)	1(x)	0(x)
0	0	1	1	0
1	1	0	1	0

The identity function x = I(x) is characterized as being at the same time an argument and a function. This distinguishes it from the three other functions of one argument, for which the value of the argument and value of the function can differ; they are opposite for the negation function ⌉(x), and invariant for the constants 1(x) and 0(x).

It is important to emphasize that the set of four Boolean functions of one argument is redundant; two of them can be expressed via the others:

I(x) = ⌉⌉(x) ,
0(x) = ⌉1(x) .

Consequently, only two of them, for example, the negation function and the logic constant 1(x), are needed to generate the rest of the set.

Boolean functions of two arguments

With two logic arguments x and y, we can construct 16 different Boolean functions The table below presents them explicitly:

x	y	↔	⊕	∧	∨	→	←	\	/	↑	↓	⊥	T	⊣	⊢	<1>	<0>
0	0	1	0	0	0	1	1	0	0	1	1	0	1	0	1	1	0
0	1	0	1	0	1	1	0	0	1	0	1	0	1	1	0	1	0
1	0	0	1	0	1	0	1	1	0	0	1	1	0	0	1	1	0
1	1	1	0	1	1	1	1	0	0	0	0	1	0	1	0	1	0

Table 2.1 Boolean functions of two arguments

The logic functions we consider are characterized by being applicable to two arguments in a certain order. All such ordered pairs constitute the range of the function. When the values of the pair are specified, a logic function yields one of two Boolean values.

With each logic function we associate a converse function, in which the order of arguments is reversed. A logic function for which

$$L(x,y) = L(y,x)$$

is called symmetric; otherwise it is asymmetric.

Under these criteria the set of 16 logic functions is divided into an octet of symmetric and an octet of asymmetric functions:

$\{\leftrightarrow, \oplus, \wedge, \vee, \uparrow, \downarrow, \hat{1}, \hat{0}\}$ – symmetric,

$\{\rightarrow, \leftarrow, \setminus, /, \perp, \top, \dashv, \vdash\}$ – asymmetric.

Of the 16 logic functions, six functions have properties which closely resemble the properties of the corresponding functions of one argument:

$$\perp(x,y) = x \perp y = x$$
$$\top(x,y) = x \top y = \bar{x}$$
$$\dashv(x,y) = x \dashv y = y$$
$$\vdash(x,y) = x \vdash y = \bar{y}$$
$$\langle 1\rangle(x,y) = x\langle 1\rangle y = 1$$
$$\langle 0\rangle(x,y) = x\langle 0\rangle y = 0$$

The reason is that these functions are actually dependent on one argument only; either x:

$$\perp(x,y) = x$$
$$\top(x,y) = \bar{x}\ ,$$

or y:

$$\dashv(x,y) = y$$
$$\vdash(x,y) = \bar{y}\ ,$$

or, finally, on neither, thus becoming the constants

$$\langle 1\rangle \quad \text{and} \quad \langle 0\rangle\ .$$

Functions $\perp$, $\top$ and $\dashv$, $\vdash$ are converses of each other:

$$x \perp y = y \dashv x$$
$$x \top y = y \vdash x\ .$$

They behave as constants with respect to one argument and as singular negations with respect to the other. The latter aspect relates these functions to the unary constants 1(x) and 0(x), which is why they are referred to as 1-negation and 0-negation functions.

In order to distinguish between unary and binary constants 1 and 0, we enclose the latter in brackets:

$$1 = x \vee \overline{x} \; ; \qquad \langle 1\rangle = x \vee \overline{x} \vee y \vee \overline{y}$$
$$0 = x \wedge \overline{x} \; ; \qquad \langle 0\rangle = x \wedge \overline{x} \wedge y \wedge \overline{y} \, .$$

In the preceding sections of this chapter we discussed several binary connectives; their corresponding logic functions are naturally contained in Table 2.1, and we shall not repeat ourselves. Instead, let us look at the remaining set of logic functions.
The function $(x \oplus y)$ we have chosen to describe as addition mod2. However, this logic function can also be interpreted as inequivalence $(x \neq y)$: indeed, the value of the function is 1 when the value of both arguments are different, and it is 0 when their values are the same. On the other hand, exactly the same requirements are assigned to exclusive disjunction: the latter yields 1 when either one or the other argument, but not both, equals 1. Consequently, we are able to give three different interpretations to the same logic function:

$$(x \oplus y) = (x \neq y) = (x \dot{\vee} y) \, .$$

It is important to be aware of these different interpretations though none of them actually will be used in the following chapter, where yet another interpretation of the same logic function will be introduced.
Function $(x \setminus y)$ is called inhibition because of the role it plays in engineering applications. The value of the function is equal to the value of the first argument when the value of the second is 0. When the second argument is 1, inhibition yields the value 0 independently of the value obtained by the first argument. For converse inhibition (x / y) these relationships are reversed.

Finally, we shall consider the function NAND $\downarrow$ and NOR $\uparrow$ which play a unique role in conventional logic. Though the connectives corresponding to these functions can easily be found in ordinary language, their discovery is due to the formal development of symbolic logic and is not the result of the formalization of general logical discourse, as was the case with the connectives AND and OR.
Function NAND, an abbreviation for Not-AND, corresponds in ordinary language to the connective "not both ..."; it yields 0 if and only if both arguments have the value 1. Function NOR, an abbreviation for Not-OR, corresponds to the connective "neither ... nor ..."; it yields 1 if and only if both arguments have the value 0. Sometimes NAND and NOR are denoted by the compound symbols which directly reflect their meaning: NAND as $\overline{\wedge}$, and NOR as $\overline{\vee}$. In this book we will not make use of these notations; instead we use the symbols specified above. The pair $\downarrow$ and $\uparrow$ is complementary to the pair $\wedge$ and $\vee$, but there is a great difference between the functions NAND and NOR, and AND and OR, which we will consider in greater detail shortly.

Basis sets

Given a set of functions one is usually interested in the possibility of expressing some functions via the others and determining the minimal irreducible set of functions – the basis set. The Boolean functions are no exception and the question naturally arises: what are the basis sets of functions which generate the complete set of logic functions?
The question of the minimal functionally complete set of logic functions is of fundamental significance in logic, both theoretically and with regard to applications.
The set of Boolean functions is redundant, and its reduction leads to interesting outcomes. In order to diminish redundancy we may begin by noticing that half of the set can be obtained from the other half by negation. Indeed,

$$\langle 0\rangle = \overline{\langle 1\rangle}$$

$$x \oplus y = \overline{x \leftrightarrow y}$$

$$x \setminus y = \overline{x \rightarrow y}$$

$$x / y = \overline{x \leftarrow y}$$

$$x \downarrow y = \overline{x \wedge y}$$

$$x \uparrow y = \overline{x \vee y}$$

$$x \perp y = \overline{x \top y}$$

$$x \dashv y = \overline{x \vdash y}$$

Hence, the operations $\langle 0\rangle$, $\oplus$, $\setminus$, $/$, $\downarrow$, $\uparrow$, $\perp$, $\dashv$ can be excluded as redundant. Taking into account that

$$x \perp y = \bar{x}$$
$$x \dashv y = \bar{y}$$

we can exclude two more redundant logic operations, thus arriving at the following set of logic functions:

constant	1
negation	$\bar{x}$
conjunction	$x \wedge y$
disjunction	$x \vee y$
implication	$x \rightarrow y$
converse implication	$x \leftarrow y$
equivalence	$x \leftrightarrow y$

Noting that converse implication is expressible by implication and negation,

$$x \leftarrow y = \bar{x} \rightarrow \bar{y}$$

we get a set of six logic functions, commonly used in conventional logic as the standard basis set. The standard set $\{1, \neg, \wedge, \vee, \rightarrow, \leftrightarrow\}$ is sufficient to express any logic function. However, further reductions are possible. Excluding the constant 1, implication and equivalence by equalities:

$$1 = x \vee \bar{x}$$
$$x \rightarrow y = \bar{x} \vee y$$
$$x \leftrightarrow y = (\bar{x} \vee y) \wedge (x \vee \bar{y})$$

we arrive at a set of elementary logic functions:

$$\{\bar{x}, x \wedge y, x \vee y\} ,$$

which contains only three connectives; NOT, AND and OR:

$$\{\neg, \wedge, \vee\} .$$

This set is the major basis set most frequently used in applications.

Minor basis sets

The major basis set allows the reduction of 20 Boolean operations, 4 unary and 16 binary, to just three functionally complete logic functions. Nevertheless, it has been discovered that further reduction of the major set is possible. Several functionally complete minor basis sets, containing only two logic functions, can be derived with the help of the De Morgan relations:

$$x \vee y = \overline{\bar{x} \wedge \bar{y}}$$
$$x \wedge y = \overline{\bar{x} \vee \bar{y}}$$

and the relation between implication $\rightarrow$ and conjunction $\wedge$ and disjunction $\vee$:

$$x \vee y = \bar{x} \rightarrow y$$
$$x \wedge y = \overline{\bar{x} \wedge y}$$

In view of the given equalities any pair of logic functions:

$$\{\rceil,\wedge\},\ \{\rceil,\vee\},\ \{\rceil,\rightarrow\}$$

represents a functionally complete minor basis set.

Minimal basis sets

The minor basis sets can be further reduced to the minimal basis sets, which implies that connectives exist that are functionally complete in themselves. Of the 20 Boolean connectives two have been discovered to possess this interesting property: NAND $\downarrow$ and NOR $\uparrow$.
Since negation and conjunction, negation and disjunction, and negation and implication have already been shown to be functionally complete, to prove that NAND or NOR are functionally complete by themselves is to show that the minor basis sets can be expressed in the unit set of NAND or NOR. We could achieve this aim by converting an expression written in one of minor basis sets into an expression written in NAND or NOR. With the help of the following equalities:

$$\bar{x} = x \downarrow x \qquad \bar{x} = x \uparrow x$$
$$x \wedge y = \overline{x \downarrow y} \qquad x \vee y = \overline{x \uparrow y}$$
$$x \vee y = \bar{x} \downarrow \bar{y} \qquad x \wedge y = \bar{x} \uparrow \bar{y}$$

the minor basis set $\{\rceil,\wedge\}$ will thus be reduced to the minimal basis set $\{\downarrow\}$ or $\{\uparrow\}$:

$$\bar{x} = (x \downarrow x) \qquad \qquad \bar{x} = (x \uparrow x)$$
$$\text{or:}$$
$$(x \wedge y) = (x \downarrow y) \downarrow (x \downarrow y) \qquad \qquad (x \wedge y) = (x \uparrow x) \uparrow (y \uparrow y)$$

Similarly, the minor basis set $\{\rceil,\vee\}$ is reducible to the minimal basis sets:

$$\bar{x} = (x \downarrow x) \qquad \qquad \bar{x} = (x \uparrow x)$$
$$\text{or:}$$
$$(x \vee y) = (x \downarrow x) \downarrow (y \downarrow y) \qquad \qquad (x \vee y) = (x \uparrow y) \uparrow (x \uparrow y)$$

The minimal basis sets $\{\downarrow\}$ and $\{\uparrow\}$ allow maximum economy in connectives to be achieved.

Unfortunately the economy in connectives is gained at the expense of clarity, for the use of a single connective can give rise to rather complicated expressions. For example, a simple logic expression like

$$x \to x$$

can, in the minimal basis set $\{\downarrow\}$, take the form:

$$((x \downarrow (x \downarrow x)) \downarrow (x \downarrow (x \downarrow x))) \downarrow ((x \downarrow (x \downarrow x)) \downarrow (x \downarrow (x \downarrow x))).$$

Clearly, in applications it is much more convenient to use major or minor basis sets, because loss of clarity is the price which must usually be paid if minimal basis sets are used. Theoretically, however, the possibility of expressing any logic function by a single connective represents a fundamental fact.

Boolean algebra

Since different logic expressions can denote the same logic function the question frequently arises whether two expressions represent the same Boolean function. The equivalence of logic functions can be verified by comparing their truth tables. This method, however, does not allow one to obtain new equalities and becomes very difficult as the number of variables increases.

The fact that the set of functions which consists of NOT, AND and OR can be used to define the complete set of Boolean functions led to the creation of the Boolean logic algebra. The latter is based on the following system of rules:

$$\overline{\overline{x}} = x$$

$x \wedge x = x$	$x \vee x = x$
$x \wedge \overline{x} = 0$	$x \vee \overline{x} = 1$
$x \wedge 1 = x$	$x \vee 1 = 1$
$x \wedge 0 = 0$	$x \vee 0 = x$
$x \wedge y = y \wedge x$	$x \vee y = y \vee x$
$x \wedge (y \wedge z) = (x \wedge y) \wedge z$	$x \vee (y \vee z) = (x \vee y) \vee z$
$x \wedge (y \vee z) = (x \wedge y) \vee (x \wedge z)$	$x \vee (y \wedge z) = (x \vee y) \wedge (x \vee z)$
$\overline{x \wedge y} = \overline{x} \vee \overline{y}$	$\overline{x \vee y} = \overline{x} \wedge \overline{y}$

The operations AND and OR of this algebra have a great deal in common with the operations of multiplication and addition in ordinary algebra, such as commutativity, associativity and distributivity of operations. However, in contrast with ordinary algebra there are not one, but two distributive laws. Furthermore, addition and multiplication of identical variables are idempotent, so that they do not yield coefficients or degrees of power.

Applying the system of equalities of the Boolean logic algebra we can now deal with various problems on a purely formal basis, and for many of them effective solutions can be formulated.

Like the arithmetic operations of ordinary algebra, logic operations in Boolean algebra have different priorities, which are reflected in the order of their execution in expressions. First, NOT is executed, followed by AND, and only then is OR performed. In cases where this order is changed, brackets are used as in algebraic expressions. In a more general case, when the standard basis set is used, brackets can be eliminated if we introduce the following ordering of connectives:

$$\leftrightarrow, \rightarrow, \vee, \wedge, \rceil.$$

This convention avoids brackets, since it is possible to restore these in the inverse order

$$\rceil, \wedge, \vee, \rightarrow, \leftrightarrow.$$

In the expression

$$\rceil x \vee y \wedge z \rightarrow y \leftrightarrow x$$

brackets will be restored as follows:

$$((((\rceil x) \vee (y \wedge z)) \rightarrow y) \leftrightarrow x).$$

However, not every Boolean expression can be written without brackets. For example, further elimination is not possible in the expressions

$$x \rightarrow (y \rightarrow z),$$
$$(x \rightarrow y) \wedge z.$$

The so-called Polish notation avoids brackets altogether. In this, the letters N, K, A, C, E (R, D) are used in the roles of negation, conjunction, disjunction, implication and equivalence, (inequivalence and NAND) respectively. The letter is written in front of variables with which it is used, hence no brackets or other notations to show association are necessary. The expression given above becomes in the Polish notation

E C A Nx Kyz y x .

The notation so obtained is unfamiliar and is not widely used. For our further discussion, however, the bracket-free Polish notation is worthy of special attention, since it contains the interesting idea of positioning logic connectives in front.

The duality principle

An important principle, which reduces by half the effort involved in the derivation of logic equalities, is the duality principle. The logic variables and logic connectives form the following dual relationships:

0 – 1
⌉ – ⌉
↔ – ⊕
∧ – ∨
↑ – ↓
→ – /
← – \
⊥ – ⊥
T – T
⊣ – ⊣
⊢ – ⊢
<0> – <1>

To form a dual formula X* from a formula X, the symbols for connectives and variables must be replaced by their duals. Then, according to the duality principle, a formula X and its dual X* are related as follows:

$$(X^*)^* = X ,$$

$$\overline{X} = X^* ,$$

$$X = \overline{X}^* .$$

If X = Y and X* is dual to X, and Y* to Y, then X* = Y*.
Examples of dual transformation can be found in the De Morgan laws:

$$\overline{x \wedge y} = \bar{x} \vee \bar{y}\,; \qquad \overline{x \vee y} = \bar{x} \wedge \bar{y}\,,$$

and in the pair of distributive laws:

$$x \wedge (y \vee z) = (x \wedge y) \vee (x \wedge z),$$

$$x \vee (y \wedge z) = (x \vee y) \wedge (x \vee z).$$

The contradiction law

$$x \wedge \bar{x} = 0$$

is dual to the law of the excluded middle:

$$\bar{x} \vee x = 1\,.$$

The duality principle, together with the algebraic transformations of Boolean algebra, enable us to construct new equalities without referring to the truth-tables. And, given a basis set of equalities, any equality of the system can be deduced from it by a series of equivalent transformations. The equivalent transformations of Boolean algebra apparently change the formula but not the logic function realized by this formula.

Tautologies and contradictions

Logic expressions which are true under any interpretation of their constituent variables are called tautologies. Logic expressions which are false under any interpretation of their constituent variables are called contradictions. In the resulting truth-table for tautology there are no other entries but 1, whereas contradiction yields only 0. Clearly, tautology and contradiction can be related by negation:

$$\text{Tautology} = \overline{\text{Contradiction}}\,.$$

For example, from the contradictions $\bar{x} \wedge x$, $x \leftrightarrow \bar{x}$, $x \setminus x$ one derives the tautologies:

$$\overline{\bar{x} \wedge x} = x \vee \bar{x}$$

$$\overline{x \leftrightarrow \bar{x}} = x \leftrightarrow x$$

$$\overline{x \setminus x} = x \rightarrow x\,.$$

An expression is a tautology if and only if its negation is a contradiction. If X and X → Y are tautologies, then Y is a tautology. If X ∧ Y is a tautology, then X and Y are tautologies. If X is a tautology, then X ∨ Y and Y ∨ X are tautologies. If X → Y is a tautology, then X* → Y* is a tautology. If X ↔ Y is a tautology, then X* ↔ Y* is a tautology.

An expression which contains only a connective ↔ is a tautology if and only if each variable enters an expression an even number of times.

Normal forms

The evaluation of logic formulas is further simplified by their reduction to a standard normal form. The fundamental theorem of Boolean algebra states that any logic formula can be expressed in the disjunctive normal form or conjunctive normal form, defined as follows:

An expression is said to be in disjunctive normal form if it represents a disjunction of one or more disjuncts, each of which is a continued conjunction of atomic variables or their negations:

$$\bigvee_{i=1}^{m} \left(\bigwedge_{j=1}^{n} x_{ij} \right)$$

Any expression which is not a contradiction is logically equivalent to some disjunctive normal form. An expression is said to be in conjunctive normal form if it represents a conjunction of one or more conjuncts, each of which is a continued disjunction of atomic variables or their negations:

$$\bigwedge_{i=1}^{m} \left(\bigvee_{j=1}^{n} x_{ij} \right)$$

Any expression which is not a tautology is logically equivalent to some conjunctive normal form.

A single variable or its negation can be considered as a degenerate disjunction or conjunction. Every disjunctive normal form is dual to some conjunctive normal form. To obtain a dual form we replace connectives by their duals and interchange 0 and 1.

$$\bigvee_{i=1}^{m} \left(\bigwedge_{j=1}^{n} x_{ij} \right) = \overline{\bigwedge_{i=1}^{m} \left(\bigvee_{j=1}^{n} \bar{x}_{ij} \right)}$$

Any formula is logically equivalent to some disjunctive or conjunctive normal form, and can be reduced to those, either directly from its truth-table or by using the method of equivalent transformations. The interpretation under which the full disjunctive normal form becomes true can be read from it directly. Another advantage lies in the ease with which validity is recognizable.

Hypotheses and consequents

Disjunctive and conjunctive normal forms are used to solve the problems of finding all the hypotheses and consequents of a given formula. Formula X is said to be a hypothesis for formula Y if

$$X \rightarrow Y = 1,$$

and formula Z is a consequent of formula Y if

$$Y \rightarrow Z = 1.$$

A hypothesis X of formula Y is called simple if it is a conjunction of atomic variables or their negations and ceases to be a hypothesis of formula Y after deletion of any of the conjuncts. Similarly, a consequent Z of formula Y is called simple if it is a disjunction of atomic variables or their negations and ceases to be a consequent after deletion of any of the disjuncts.

The solution to the problem of finding hypotheses and consequents requires the construction of all the simple hypotheses and consequents for a given formula, and then obtaining all other hypotheses and consequents with the help of equivalent transformations.

A disjunct of a disjunctive normal form is a hypothesis of this disjunctive normal form; a conjunct of a conjunctive normal form is a consequent of this disjunctive normal form.

A full disjunctive normal form has no other hypothesis but the disjunctions of some of its disjuncts. A full conjunctive normal form has no other consequents but the conjunction of some of its conjuncts.

A reduced disjunctive normal form is a disjunction of its simple hypotheses; a reduced conjunctive normal form is a conjunction of its simple consequents.

A reduced disjunctive normal form has important applications for the minimization of logic functions. The minimization of logic functions requires the construction of a disjunctive normal form which has a minimal number of conjuncts in its disjuncts, and thus minimal complexity. Such a disjunctive normal form is called a minimal disjunctive normal form.

2.2 Higher-order logics

The first-order Boolean logic discussed in the preceding section covers aspects of logic that are of great theoretical and applied significance. However, there are logical truths that lie wholly beyond the grasp of first-order logic. A broader analysis of the inner structure of logic expressions is achieved in such higher-order logics as predicate logic and modal logics.

Predicate logic

In predicate or quantificational logic, a logic function is an expression that becomes a statement when the variables involved are replaced by particular predicates and particular terms. The basic idea behind predicates, or expressions that yield statements when attached to particular terms, is the formation of higher-order logic functions. The n-ary predicate $P(x_1, x_2, \ldots x_n)$ itself is not a proposition, but when particular terms are substituted for the variables, the predicate becomes a statement, either true or false. If m variables, $m < n$, are assigned particular terms, the n-ary predicate is changed into an (n-m) - ary predicate. From this point of view, the expressions of first-order logic can be considered as 0-ary predicates, in which all variables are substituted by particular terms.

Given a set of predicates, and making use of Boolean connectives, we can construct new predicates. In addition to Boolean connectives, predicate logic also makes use of quantifiers, which immensely enlarge its expressive power. Quantifiers, which change a predicate P(x) into a particular statement, are commonly of two kinds:

The existential quantifier $\exists$ P(x), reads: "there exists an x such that P(x)", and the universal quantifier $\forall$ xP(x), reads: "for every x, P(x)". Quantification $\exists$ xP(x) is true if and only if P(x) is true at least for one x from the domain of discourse $\{x_n\}$.

Quantification $\forall$ xP(x) is true if and only if P(x) is true for all x from the domain of discourse $\{x_n\}$. Thus, the existential quantifier generalizes the operation of disjunction in the same way as the universal quantifier generalizes the operation of conjunction:

$$\exists P(x) = P(x_1) \vee P(x_2) \vee P(x_3) \vee \ldots \vee P(x_n)$$

$$\forall P(x) = P(x_1) \wedge P(x_2) \wedge P(x_3) \wedge \dots \wedge P(x_n)$$

Quantifiers can be used in various complex ways and appear in logic expressions which contain both bound occurences of variables, which are governed by a quantifier, and free occurrences, which are not.

The central theorem of predicate logic states that any logic expression containing variables, logic connectives and quantifiers can be brought into normal form such that all the quantifiers are carried out into the front of the expression:

$$\exists x_1 \exists x_2 \dots \exists x_i \forall x_1 \forall x_2 \dots \forall x_j P(x_1\ x_2\ x_3 \dots x_n).$$

In most formalizations the existential and universal quantifiers are considered to be basic and the other quantifiers are explained in terms of them. Moreover, the existential and universal quantifiers are actually not both necessary, if negation is available.

The rules for an exchange of quantifiers generalizes the duality law. Since $\neg\exists x \neg P(x)$ denies that there is x such that P(x) fails to hold true, it makes the same assertion as $\forall xP(x)$, namely P(x) holds true for all x:

$$\neg\exists x \neg P(x) = \forall xP(x).$$

Similarly, $\neg\forall x \neg P(x)$ makes precisely the same assertion as $\exists xP(x)$:

$$\neg\forall x \neg P(x) = \exists xP(x).$$

These equivalences have a general meaning independent of a particular form of the function P(x). Consequently, we have the following two fundamental logical constructs:

$$\neg\exists\neg = \forall,$$
$$\neg\forall\neg = \exists$$

where bilateral application of negation converts a quantifier into its corresponding counterpart.

Modal logic

The vocabulary of conventional logic is further expanded in modal logic, which deals with the concepts of necessity and possibility. It would be quite impossible to present here the results of modal logic in full, and that is not our intention. We would like to emphasize one important aspect of modal logic relevant to our further discussion. When dealing with ordinary Boolean expressions we apparently are interested in their actual truth-values. As far as the truth-value of an expression is concerned, only the actual truth-value of an expression is relevant; its possible truth-value outcomes are quite immaterial.

The introduction of modal connectives, $\Box x$, read as 'necessarily x; and $\Diamond x$, read as 'possibly x', shows that the value of a logic expression can depend on more than the actual truth-values of its components. In modal logic, the truth-value of expressions can also depend on the possible truth-value outcomes. If truth is a possible truth-value outcome of x then $\Diamond x$ is true; otherwise $\Diamond x$ is false.
The relationship between the modal connectives $\Box$ and $\Diamond$ is similar to the fundamental relationship between the universal $\forall$ and existential $\exists$ quantifiers. Since necessarily $\bar{x}$, $\Box\,\bar{x}$, makes precisely the same assertion as impossible x, $\neg\Diamond x$, whereas not necessarily x, $\neg\Box x$, is equivalent to possibly $\bar{x}$, $\Diamond\,\bar{x}$, we can deduce the following fundamental relationships between the modal connectives:

$$\Box = \neg \Diamond \neg$$

$$\Diamond = \neg \Box \neg$$

Just as in the relation between existential and universal quantifiers, the connectives $\Box$ and $\Diamond$ are interchangeable and the expressive power of modal logic is in no way impaired by the omission of either of them.

2.3 Multivalued logic

In multivalued logics, pioneered by J. Lucasiewicz and E. Post, the class of logic functions is expanded to incorporate those functions whose arguments, as well as the functions themselves, are defined in more than a binary domain. It is possible to distinguish two related multivalued approaches. In one of them the addition of truth-values encodes intermediate levels of validity between true or valid and false or invalid. In another probabilistic approach the addition of truth-values encodes intermediate levels of certainty between complete information and incomplete information. Clearly, the introduction of multivalued logic systems is possible only at the expense of rejecting the universal validity of the classical law of the excluded middle:

$$x \vee \bar{x} = 1,$$

which gives rise to ternary or three-valued logic; the excluded 4th, which gives rise to the tetrary or four-valued logic, or, generally, the excluded kth which allows one to expand (k–1)-valued logic into a k-valued logic.

In expanding the alphabet of truth-values, multivalued logic represents a generalization of Boolean two-valued logic. The multivalued logic funtions are defined on the set

$$E_k = \{0, 1, \dots, k-1\} ; \quad k = 2, 3, \dots$$

which for k = 2 becomes the binary set $E_2 = \{0,1\}$ of two-valued logic. The number of k-valued functions of n arguments $x_1, x_2, \dots x_n$ is k^{k^n}. For $k \geq 3$, the difficulties of truth-table presentation and effective enumeration of all possible functions, in comparison with binary logic, increase rapidly. In three-valued logic there are $3^9 = 19683$ possible truth functions with two arguments, a significant difference from two-valued logic, where there are only 16 different functions of two arguments.

It is quite understandable that multivalued logic systems have been attracting growing attention for their wealth of logic structures. The elementary functions of binary logic have several analogues in multivalued systems. Consider some generalizations of negation:

$\bar{x} = x + 1 \pmod{k}$ – cyclic negation,

$\sim x = k - 1 - x$ – diametrical negation.

Cyclic negation is a generalization of negation involving a cyclic shift of values. Another aspect of negation is generalized by diametrical negation, which is general multivalued complementation.

There are also several generalizations for the logic functions of two arguments. The most commonly used generalizations of conjunction and disjunction are the following:

Conjunction	Disjunction
a) min (x,y)	a) max (x,y)
b) x . y (mod k)	b) x + y (mod k) .

The multivalued functions in many respects resemble the functions of binary logic and some results of the latter remain applicable in multivalued logic. However, there are important novel properties which distinguish a logic for $k \geq 3$ from classical binary logic. Consider (for $k \geq 3$):

$$\sim (\sim x) = x \qquad \text{but} \qquad \bar{\bar{x}} \neq x$$

$$\sim \min (x,y) = \max (\sim x, \sim y) \qquad \text{but} \qquad \overline{\min (x,y)} \neq \max (\bar{x}, \bar{y}) \; .$$

An important class of multivalued functions is represented by the Post functions:

$$J_i(x) = \begin{cases} k-1 & \text{for} \quad x = i \\ 0 & \text{for} \quad x \neq i. \end{cases}$$

$$i = 0, k-1.$$

The functions $J_i(x)$ allow reduction of multivalued functions to normal forms, the analogue of fundamental reduction in binary Boolean logic to disjunctive normal forms.

The question of the relation between binary and multivalued logics is not as simple as it may appear. The most direct intuitive perception might be that binary logic is a particular reduction of multivalued logic which reinstates the law of the excluded middle. On the other hand, proposals have been made to express the functions of multivalued logic in the binary logarithmic code $\log_2 k$. This is not only useful for computer science applications, but also casts new light on fundamentals.

An intuitive assumption that multivalued logic is more general than binary and contains the latter as a degenerate case runs counter to our understanding of the role of classical binary logic, the logic we actually use to describe multivalued logics.

2.3 Quantum logic

Quantum logic represents one of most recent developments in conventional logic.Its appearance followed the realization that the logical problems arising in quantum physics cannot be accommodated by Boolean logic alone.

The logic underlying quantum mechanics could not be a Boolean algebra (lattice), since it must admit propositions that can be neither true nor false.

Originating with a paper by G. Birkhoff and J. van Neumann in 1936, quantum logic denies the universal validity of the commutative and distributive laws of classical logic. In the general case:

$$x \wedge y \neq y \wedge x$$
$$x \vee y \neq y \vee x$$
$$x \wedge (y \vee z) \neq (x \wedge y) \vee (x \wedge z)$$
$$x \vee (y \wedge z) \neq (x \vee y) \wedge (x \vee z)$$

This revision of classical logic is adopted in order to accommodate the behaviour of quantum fields and particles, which display properties that do not conform to the classical rules of inference. Clearly, in the quantum domain the logic relationships are altered profoundly.

A commutation relation defines a fundamental limit to observation and one reasonably expects classical logic to fail in the quantum domain whenever the logic variables in an argument do not permute.

In the case of the distributive laws, even though the left sides of corresponding equations may be true, the right sides cannot be satisfied. Quantum logic succeeds in retaining the logical validity of quantum theory but at the expense of revising the roles of the connectives $\wedge$ and $\vee$, which are altered so that compliance with the distributive laws is no longer the universal requirement for logic relations.

The three-valued approach to quantum logic proposed by H. Reichenbach in 1944 introduces the indeterminate truth-value in addition to true and false. Instead of focusing on the distributive laws Reichenbach defines enriched patterns of logical connectives by means of truth-tables. With the third truth-value added to the set of truth-values, the truth-tables for $\wedge$ and $\vee$ take the following form, where "i" denotes the indeterminate truth-value:

x	y	$x \wedge y$	$x \vee y$
1	1	1	1
1	i	i	1
1	0	0	1
i	1	i	1
i	i	i	i
i	0	0	i
0	1	0	1
0	i	0	i
0	0	0	0

An attractive feature of this approach is that the truth-tables of quantum-logical conjunction and disjunction, with deletion of the indeterminate truth-value i, are reduced to the corresponding classical truth-tables.

Relieved of classical constraints, quantum logic proved to be useful in the domain of quantum-mechanical phenomena. However, a debate is still going on whether it should be considered as a genuinely new logic or merely as an algebra under another name. One extreme is to view it not only as a new logic but as one even more fundamental than classical logic. The opposite view gives priority to classical logic, since quantum logic is rooted in quantum mechanics, which as a theory makes extensive use of classical logic.

Thus it seems a contradiction in terms that this quantum theory can give rise to a revision of logic which denies classical logic itself. This problem bears a resemblance to the relation between binary and multivalued logic, but as no other development, quantum logic has brought into focus the question whether logic can be derived from some fundamental base or whether there are truths of logic that would hold no matter what the condition of the world.

Introspective logical considerations alone cannot resolve this question. Both the introspective design of logic, which mirrors in its structure the properties of human mind (Hume's relations of ideas), and quantum logic design, which seeks to satisfy the oddities of experimental physics (Hume's matters of fact), are constructed heuristically and lack the convincing feature of derivability from some objective foundation.

Should we change logic theory every time new facts of physics are discovered? Could logic be derived from some universal fundamental principles? Quantum logic has considerably reshaped our view of logic and brought the understanding that an adequate theory of logic must model itself on the exact sciences.

CHAPTER 3

MATRIX FORMULATION OF LOGIC

Classical logic was devised to articulate and communicate logical truths. The use of symbolic notation and the development of Boolean logic algebra has brought logic closer to the realm of mathematics and the exact sciences.
However, in spite of being able to capture important formal aspects of deductive reasoning, logic still lacks the decisive features that would allow it to function fully as an instrument of calculation. In this chapter we enlarge the computational capabilities of logic by interpreting logic connectives as matrix operators acting in two adjoining spaces of logic vectors.

3.1 Positive logic Ω^+

As a first step towards a matrix formulation of logic we shall regroup the truth-tables of the binary Boolean functions and present them in the form of square tables:

Equivalence $\leftrightarrow$

x \ y	0	1
0	1	0
1	0	1

Implication $\rightarrow$

x \ y	0	1
0	1	1
1	0	1

Inequivalence $\neq$

x \ y	0	1
0	0	1
1	1	0

Converse implication $\leftarrow$

x \ y	0	1
0	1	0
1	1	1

Conjunction ∧

x \ y	0	1
0	0	0
1	0	1

Nand ↓

x \ y	0	1
0	1	1
1	1	0

Disjunction ∨

x \ y	0	1
0	0	1
1	1	1

Nor ↑

x \ y	0	1
0	1	0
1	0	0

Inhibition \

x \ y	0	1
0	0	0
1	1	0

Converse inhibition /

x \ y	0	1
0	0	1
1	0	0

1-negation ⊣

x \ y	0	1
0	0	1
1	0	1

0-negation ⊢

x \ y	0	1
0	1	0
1	1	0

Converse 1-negation $\perp$

x \ y	0	1
0	0	0
1	1	1

Converse 0-negation T

x \ y	0	1
0	1	1
1	0	0

Tautology constant $\hat{1}$

x \ y	0	1
0	1	1
1	1	1

Empty operator $\hat{0}$

x \ y	0	1
0	0	0
1	0	0

Clearly, the information contained in the square representations of the Boolean functions is precisely the same as we had in the column representation in Chapter 2. The only difference is that now each table has its own reference variables: x is indicated in the leftmost column, and y in the topmost row. The four different values inside the tables are the values which are yielded by the corresponding Boolean function, provided that the pair of variables x and y are given a specific assignment, namely {0,0}, {0,1}, {1,0}, {1,1}.

Operator postulate

Matrix logic is founded on the following operator postulate: instead of viewing the given truth-tables as a convenient illustration introduced simply for the purpose of friendly usage, we will interpret the tables as mathematical objects, namely as (2×2)-matrices defined as follows:

Name	Symbol	L	$\tilde{L}$
Equivalence	$\leftrightarrow$	$\begin{pmatrix} 1 & 0 \\ 0 & 1 \end{pmatrix}$	$\begin{pmatrix} 1 & 0 \\ 0 & 1 \end{pmatrix}$
Negation	$\urcorner$	$\begin{pmatrix} 0 & 1 \\ 1 & 0 \end{pmatrix}$	$\begin{pmatrix} 0 & 1 \\ 1 & 0 \end{pmatrix}$
Conjunction	$\wedge$	$\begin{pmatrix} 0 & 0 \\ 0 & 1 \end{pmatrix}$	$\begin{pmatrix} 0 & 0 \\ 0 & 1 \end{pmatrix}$
Disjunction	$\vee$	$\begin{pmatrix} 0 & 1 \\ 1 & 1 \end{pmatrix}$	$\begin{pmatrix} 0 & 1 \\ 1 & 1 \end{pmatrix}$
Implication	$\rightarrow$	$\begin{pmatrix} 1 & 1 \\ 0 & 1 \end{pmatrix}$	$\begin{pmatrix} 1 & 0 \\ 1 & 1 \end{pmatrix}$
Converse implication	$\leftarrow$	$\begin{pmatrix} 1 & 0 \\ 1 & 1 \end{pmatrix}$	$\begin{pmatrix} 1 & 1 \\ 0 & 1 \end{pmatrix}$
Inhibition	$\backslash$	$\begin{pmatrix} 0 & 0 \\ 1 & 0 \end{pmatrix}$	$\begin{pmatrix} 0 & 1 \\ 0 & 0 \end{pmatrix}$
Converse inhibition	$/$	$\begin{pmatrix} 0 & 1 \\ 0 & 0 \end{pmatrix}$	$\begin{pmatrix} 0 & 0 \\ 1 & 0 \end{pmatrix}$
Nor	$\uparrow$	$\begin{pmatrix} 1 & 0 \\ 0 & 0 \end{pmatrix}$	$\begin{pmatrix} 1 & 0 \\ 0 & 0 \end{pmatrix}$
Nand	$\downarrow$	$\begin{pmatrix} 1 & 1 \\ 1 & 0 \end{pmatrix}$	$\begin{pmatrix} 1 & 1 \\ 1 & 0 \end{pmatrix}$
1-negation	$\dashv$	$\begin{pmatrix} 0 & 1 \\ 0 & 1 \end{pmatrix}$	$\begin{pmatrix} 0 & 0 \\ 1 & 1 \end{pmatrix}$
0-negation	$\vdash$	$\begin{pmatrix} 1 & 0 \\ 1 & 0 \end{pmatrix}$	$\begin{pmatrix} 1 & 1 \\ 0 & 0 \end{pmatrix}$
Converse 1-negation	$\perp$	$\begin{pmatrix} 0 & 0 \\ 1 & 1 \end{pmatrix}$	$\begin{pmatrix} 0 & 1 \\ 0 & 1 \end{pmatrix}$
Converse 0-negation	$\top$	$\begin{pmatrix} 1 & 1 \\ 0 & 0 \end{pmatrix}$	$\begin{pmatrix} 1 & 0 \\ 1 & 0 \end{pmatrix}$
Tautology	$\hat{1}$	$\begin{pmatrix} 1 & 1 \\ 1 & 1 \end{pmatrix}$	$\begin{pmatrix} 1 & 1 \\ 1 & 1 \end{pmatrix}$
Empty operator	$\hat{0}$	$\begin{pmatrix} 0 & 0 \\ 0 & 0 \end{pmatrix}$	$\begin{pmatrix} 0 & 0 \\ 0 & 0 \end{pmatrix}$

Notice one significant change: the matrix corresponding to inequivalence (the same exclusive disjunction or addition mod2) we denoted by a symbol $\rceil$, previously used for denoting negation. The meaning of this innovation will be clarified shortly.

Treating logic connectives as matrices allows one to explore effectively the operation of matrix transposition which provides a mathematical foundation for what in conventional logic is known as the converse logic function.

The octet of logic connectives $\{\leftrightarrow, \rceil, \wedge, \dot{\vee}, \uparrow, \downarrow, \hat{1}, \hat{0}\}$ is represented by symmetric matrices and consequently, the connectives and their converses are the same. The remaining octet is subdivided into two sets which are converses of each other, or in matrix logic terms, matrix transposes of each other.

Connectives as matrix operators

We shall interpret the truth matrices defined above as (2×2) matrix operators and refer to these as logic operators or in short, L-operators. In developing this line of reasoning one should of course define an appropriate vector space in which logic operators can act as such. The vector space formalism indeed offers a concise, consistent notation which is ideally suited to the needs of matrix logic.

Let us introduce two adjoint spaces of logic vectors: a row vector space consisting of the bra-vectors $\langle x|$ and a column vector space consisting of the ket-vectors $|y\rangle$.

To be compatible with L-operators the bra and ket vector spaces are, naturally, 2-dimensional:

$$\langle x| = (x_1, x_2)$$

$$|y\rangle = \begin{pmatrix} y_1 \\ y_2 \end{pmatrix}$$

$$x_i \in E_2 = \{0,1\}$$

$$y_j \in E_2 = \{0,1\}$$

We consider vector spaces inhibited by the vectors of the unit norm

$$\langle x \mid x\rangle = \langle y \mid y\rangle = 1$$

with the inner product of two vectors defined as

$$<x \mid y> = \delta_{xy} = \begin{cases} 1 & \text{if } x = y \\ 0 & \text{if } x \neq y \end{cases}$$

where δ_{xy} is the Kronecker delta function.
This implies that

$$<x \mid y> \in E_2 = \{0,1\}.$$

In order to be able to compare the results of matrix logic with those of conventional logic, we put in correspondence to the Boolean scalars 1 and 0, the logic vectors (0,1) and (1,0) (bra or ket) respectively.

With the transparent definitions just described, we can now generate 16 binary Boolean functions by employing one fundamental device which is a bilinear product defined on the binary set $E_2 = \{0,1\}$.

$$<x \mid L \mid y> = x_1 L_{11} y_1 + x_1 L_{12} y_2 + x_2 L_{21} y_1 + x_2 L_{22} y_2$$

where L_{11}, L_{12}, L_{21}, L_{22} are matrix elements of a logic operator L, and x_1, x_2, y_1, y_2 are components of the bra and ket logic vectors $<x|$ and $|y>$.
The concrete value of a logic function is now obtained by means of one fundamental device of matrix multiplication, and not via the tedious procedure of referring to a corresponding square of truth-tables.
Consider the following matrix multiplications, which are due to $<x \mid \wedge \mid y>$:

$$(1\ \ 0)\begin{pmatrix} 0 & 0 \\ 0 & 1 \end{pmatrix}\begin{pmatrix} 1 \\ 0 \end{pmatrix} = (0\ \ 0)\begin{pmatrix} 1 \\ 0 \end{pmatrix} = (1\ \ 0)\begin{pmatrix} 0 \\ 0 \end{pmatrix} = 0$$

$$(1\ \ 0)\begin{pmatrix} 0 & 0 \\ 0 & 1 \end{pmatrix}\begin{pmatrix} 0 \\ 1 \end{pmatrix} = (0\ \ 0)\begin{pmatrix} 0 \\ 1 \end{pmatrix} = (1\ \ 0)\begin{pmatrix} 0 \\ 1 \end{pmatrix} = 0$$

$$(0\ \ 1)\begin{pmatrix} 0 & 0 \\ 0 & 1 \end{pmatrix}\begin{pmatrix} 1 \\ 0 \end{pmatrix} = (0\ \ 1)\begin{pmatrix} 1 \\ 0 \end{pmatrix} = (0\ \ 1)\begin{pmatrix} 0 \\ 0 \end{pmatrix} = 0$$

$$(0\ \ 1)\begin{pmatrix} 0 & 0 \\ 0 & 1 \end{pmatrix}\begin{pmatrix} 0 \\ 1 \end{pmatrix} = (0\ \ 1)\begin{pmatrix} 0 \\ 1 \end{pmatrix} = (0\ \ 1)\begin{pmatrix} 0 \\ 1 \end{pmatrix} = 1$$

The logic operator $\vee$ would produce other results for $\langle x | \vee | y \rangle$ which correspond fully to conventional conjunction:

$$(1\ \ 0)\begin{pmatrix}0 & 1\\ 1 & 1\end{pmatrix}\begin{pmatrix}1\\ 0\end{pmatrix} = (0\ \ 1)\begin{pmatrix}1\\ 0\end{pmatrix} = (1\ \ 0)\begin{pmatrix}0\\ 1\end{pmatrix} = 0$$

$$(1\ \ 0)\begin{pmatrix}0 & 1\\ 1 & 1\end{pmatrix}\begin{pmatrix}0\\ 1\end{pmatrix} = (0\ \ 1)\begin{pmatrix}0\\ 1\end{pmatrix} = (1\ \ 0)\begin{pmatrix}1\\ 1\end{pmatrix} = 1$$

$$(0\ \ 1)\begin{pmatrix}0 & 1\\ 1 & 1\end{pmatrix}\begin{pmatrix}1\\ 0\end{pmatrix} = (1\ \ 1)\begin{pmatrix}1\\ 0\end{pmatrix} = (0\ \ 1)\begin{pmatrix}0\\ 1\end{pmatrix} = 1$$

$$(0\ \ 1)\begin{pmatrix}0 & 1\\ 1 & 1\end{pmatrix}\begin{pmatrix}0\\ 1\end{pmatrix} = (1\ \ 1)\begin{pmatrix}0\\ 1\end{pmatrix} = (0\ \ 1)\begin{pmatrix}1\\ 1\end{pmatrix} = 1$$

We need not continue to illustrate the computational capability of the fundamental logic device introduced here. One can easily verify that the bilinear product $\langle x | L | y \rangle$, being inserted with a logic operator and a pair of logic vectors, will produce correct and meaningful results for all Boolean functions without exception.

Special attention must be paid to the peculiarity of the logic vector spaces: vectors here are permitted to interact only via connectives. In contrast with ordinary vector space, the addition operation for logic vectors is not defined.

However, multiplicative interactions of logic vectors are allowed in the form of the inner $\langle x | y \rangle$ and outer products $|x\rangle\langle y|$. Of these two, $\langle x | y \rangle$ is, actually, $\langle x | \leftrightarrow | y \rangle$ i.e. a case in which vectors interact via a connective operator.

Unary and binary operations

In conventional logic there is a strict division into unary and binary connectives. In matrix logic we have given the logic connectives the direct mathematical status of matrix operators, replacing the 16 binary Boolean operations with the universal procedure of matrix multiplication. It seems quite logical to consider, in addition to the binary operations, the unary operations of logic operators. In doing so, one ought to recognize that in matrix logic, as in conventional logic, a division into unary and binary operations does exist, but a division into unary and binary connectives or operators ceases to have similar implications. Indeed, in contrast to conventional connectives, the same logic operator is able to act either as a binary operator between two logic vectors, or as a unary operator upon a single logic vector. This implies, furthermore, that the set of 4 unary operations of Boolean logic is enlarged to a set of 16 unary operations, making the number of binary and unary operations in matrix logic equal, in total, 32 operations.

This analysis of the dual nature of matrix logic operators has finally brought us to the point where the reason for replacing the operator symbol $\neq$ (or $\dot{\vee}$, or $\oplus$) with the negation symbol $\urcorner$ can be meaningfully explained.

The negation operator

The binary matrix inequivalence operator, acting as a unary operator upon the logic vectors, converts them into their complementary. We have for the ket-logic vectors:

$$\begin{pmatrix} 0 & 1 \\ 1 & 0 \end{pmatrix}\begin{pmatrix} 1 \\ 0 \end{pmatrix} = \begin{pmatrix} 0 \\ 1 \end{pmatrix}$$

$$\begin{pmatrix} 0 & 1 \\ 1 & 0 \end{pmatrix}\begin{pmatrix} 0 \\ 1 \end{pmatrix} = \begin{pmatrix} 1 \\ 0 \end{pmatrix}$$

and for the bra-logic vectors:

$$(1 \quad 0)\begin{pmatrix} 0 & 1 \\ 1 & 0 \end{pmatrix} = (0 \quad 1)$$

$$(0 \quad 1)\begin{pmatrix} 0 & 1 \\ 1 & 0 \end{pmatrix} = (1 \quad 0)$$

In view of the assignments given to the logical vectors (0,1) and (1,0), the inequivalence operator demonstrates in a matrix formulation of logic the fundamental properties of negation, and will be used as such throughout the rest of this book. The negation operator $\urcorner$ is of crucial importance in matrix logic. It is also the only case of a classical connective being given a new interpretation; this adds yet another possible interpretation to a complex connective which has already been interpreted as inequivalence, exclusive disjunction and addition mod2.

Equivalence and identity

Another important case of the dual nature of matrix logic operators is the unary application of the binary operator $\leftrightarrow$. The binary application of $\leftrightarrow$ is equivalence, which in the matrix formulation takes the form of the Kronecker function:

$$<x \mid \leftrightarrow \mid y> = <x \mid y> = \delta_{xy} = \begin{cases} 1 & \text{for } x = y \\ 0 & \text{for } x = y \end{cases}$$

On the other hand the operator $\leftrightarrow$, being applied as a unary operator, represents the identity function I or affirmation discussed in the previous chapter: clearly, multiplication of a vector by a unit matrix reproduces the given vector:

$$\leftrightarrow \mid x> = I \mid x> = \mid x> ,$$

$$<x \mid \leftrightarrow = <x \mid I = <x \mid .$$

Left logic and right logic

A correspondence can be established between the two remaining unary functions of conventional logic, the logical constants 1 and 0, and the logic operators of matrix logic. If arguments are defined in the bra vector space, the operator 1-negation will convert any logic vector into a vector (0,1) - matrix representation of the Boolean 1; and 0-negation will convert any logic vector into a (1,0) - matrix representation of the Boolean 0:

$$<x \mid \dashv = <\bar{x} \mid \dashv = (0,1)$$

$$<x \mid \vdash = <\bar{x} \mid \vdash = (1,0)$$

However, if the arguments are the ket logic vectors, the operators of 1-negation and 0-negation will not do the job. To preserve consistency we are obliged to replace the operators by their converses:

$$\perp \mid x> = \perp \mid \bar{x}> = \begin{pmatrix} 0 \\ 1 \end{pmatrix}$$

$$T \mid x> = T \mid \bar{x}> = \begin{pmatrix} 1 \\ 0 \end{pmatrix}$$

This analysis illuminates the important fact that the left and right application of a logic operator can be different. By interpreting logic connectives as matrix operators and introducing two

adjoint spaces of logic vectors, we implicitly defined two sorts of logics which we shall call left logic and right logic. This new concept is due to the fundamental ability of logic operators to act either from the right (right logic) or from the left (left logic), or both (left/right logic).
The set of matrix logic operators $\{\Omega^+\}$ decays into two subsets: one of them

$$\{\Omega^+_{sym}\} = \{\leftrightarrow, \urcorner, \wedge, \vee, \uparrow, \downarrow, \hat{1}, \hat{0}\}$$

acts identically both in left and in right logics, whereas the other subset:

$$\{\Omega^+_{asym}\} = \{\rightarrow, \leftarrow, \backslash, /, \perp, \top, \dashv, \vdash\}$$

acts differently in left logic and in right logic. As indicated by the subscripts, the subset $\{\Omega^+_{sym}\}$ consists of operators whose matrices are symmetric, and subset $\{\Omega^+_{asym}\}$ of operators whose matrices are asymmetric.

Undecided in matrix logic

We intentionally began developing the formalism of matrix logic with no new premises beyond those we accept in binary Boolean logic. However, thanks to the application of the powerful mathematics of matrix calculus, matrix logic rapidly begins to unfold new and hitherto hidden aspects of logic.
Since classical logic is a two-valued logic, in which every variable is either true 1 or false 0, we have introduced their matrix equivalents, the logic vectors (0,1) and (1,0) respectively. Exploring this new notation we have not only obtained the complete set of Boolean functions in full correspondence with conventional logic, but also enlarged the number of the functions from 20 to 32, using the new dual capabilities of matrix logic operators.
In addition, we have discovered that the unary application of logic operators in some cases reveals some peculiar and unexpected properties of matrix logic. On the one hand, the unary application of such operators as $\leftrightarrow$ and $\urcorner$ (both in left and right logic), $\perp$ and $\top$ (in left logic only), $\dashv$ and $\vdash$ (in right logic only) fulfils the requirement of binary logic; the result of these applications is one of the binary logic vectors. But, on the other hand, the unary application of such operators as

$$\wedge, \vee, \rightarrow, \leftarrow, \backslash, /, \uparrow, \downarrow, \perp, \top, \dashv, \vdash, \hat{1}, \hat{0}$$

may lead to the emergence of two new logic vectors:

$$\begin{pmatrix} 0 \\ 0 \end{pmatrix} \text{ and } \begin{pmatrix} 1 \\ 1 \end{pmatrix}$$

Just as with the assignment of the value 1(true) to the vector (0,1) and the value 0(false) to the vector (1,0), we can associate the vectors (0,0) and (1,1) with the truth-value "undecided". Thus, without any particular effort, we can expand two-valued binary logic into a ternary logic of true, false and undecided, where a vector (0,0) represents weak undecided –"neither true nor false", and vector (1,1) represents strong undecided –"both true and false".
As is now apparent, the ternary logic of undecided is not a separate logic but is actually a hidden aspect of binary logic, which reveals itself in the matrix formulation.
If we wish to distinguish between (0,0) and (1,1) as two separate truth-values – minimal and maximal undecided – we may further expand the ternary logic into quaternion logic with four different truth-values, forming two complementary pairs:

$$\begin{pmatrix} \bar{0} \\ 1 \end{pmatrix} = \begin{pmatrix} 1 \\ 0 \end{pmatrix}$$

and

$$\begin{pmatrix} \bar{1} \\ 1 \end{pmatrix} = \begin{pmatrix} 0 \\ 0 \end{pmatrix}.$$

Presenting the logic vectors by means of vertices of the Boolean lattice in matrix logic we get a two-fold increase in the number of possible logic states:

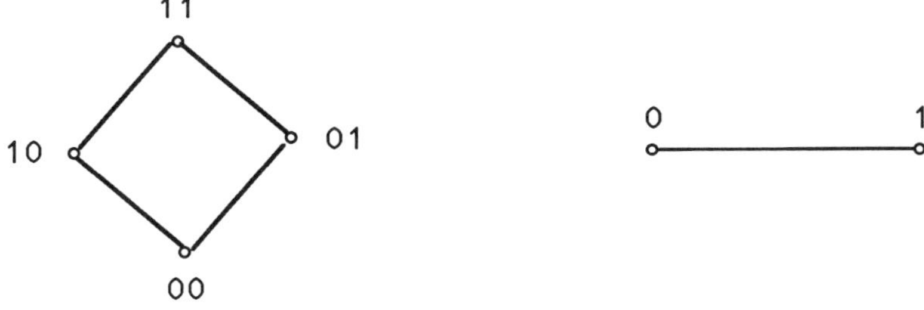

logic vectors logic scalars

Logic operators extend beyond classical logic since they deal not only with true and false but also with undecided truth-values.

In the bilinear logic products <x | L | y>, the undecided vectors appear as intermediary virtual states of computation, which do not show themselves in the final results. The undecided states appear in the following cases:

$\perp$	— —	$(0\ \ 1)\begin{pmatrix}0 & 0\\1 & 1\end{pmatrix} = (1\ \ 1)$ $(1\ \ 0)\begin{pmatrix}1 & 1\\0 & 0\end{pmatrix} = (0\ \ 0)$
$\top$	— —	$(0\ \ 1)\begin{pmatrix}1 & 1\\0 & 0\end{pmatrix} = (0\ \ 0)$ $(1\ \ 0)\begin{pmatrix}1 & 1\\0 & 0\end{pmatrix} = (1\ \ 1)$
$\dashv$	$\begin{pmatrix}0 & 1\\0 & 1\end{pmatrix}\begin{pmatrix}0\\1\end{pmatrix} = \begin{pmatrix}1\\1\end{pmatrix}$ $\begin{pmatrix}1 & 0\\1 & 0\end{pmatrix}\begin{pmatrix}1\\0\end{pmatrix} = \begin{pmatrix}0\\0\end{pmatrix}$	— —
$\vdash$	$\begin{pmatrix}1 & 0\\1 & 0\end{pmatrix}\begin{pmatrix}0\\1\end{pmatrix} = \begin{pmatrix}0\\0\end{pmatrix}$ $\begin{pmatrix}1 & 0\\1 & 0\end{pmatrix}\begin{pmatrix}1\\0\end{pmatrix} = \begin{pmatrix}1\\1\end{pmatrix}$	— —
$\hat{1}$	$\begin{pmatrix}1 & 1\\1 & 1\end{pmatrix}\begin{pmatrix}0\\1\end{pmatrix} = \begin{pmatrix}1\\1\end{pmatrix}$ $\begin{pmatrix}1 & 1\\1 & 1\end{pmatrix}\begin{pmatrix}1\\0\end{pmatrix} = \begin{pmatrix}1\\1\end{pmatrix}$	$(0\ \ 1)\begin{pmatrix}1 & 1\\1 & 1\end{pmatrix} = (1\ \ 1)$ $(1\ \ 0)\begin{pmatrix}1 & 1\\1 & 1\end{pmatrix} = (1\ \ 1)$
$\hat{0}$	$\begin{pmatrix}0 & 0\\0 & 0\end{pmatrix}\begin{pmatrix}0\\1\end{pmatrix} = \begin{pmatrix}0\\0\end{pmatrix}$ $\begin{pmatrix}0 & 0\\0 & 0\end{pmatrix}\begin{pmatrix}1\\0\end{pmatrix} = \begin{pmatrix}0\\0\end{pmatrix}$	$(0\ \ 1)\begin{pmatrix}0 & 0\\0 & 0\end{pmatrix} = (0\ \ 0)$ $(1\ \ 0)\begin{pmatrix}0 & 0\\0 & 0\end{pmatrix} = (0\ \ 0)$

$\wedge$	$\begin{pmatrix}0&0\\0&1\end{pmatrix}\overline{\begin{pmatrix}1\\0\end{pmatrix}} = \begin{pmatrix}0\\0\end{pmatrix}$	$(1\quad 0)\overline{\begin{pmatrix}0&0\\0&1\end{pmatrix}} = (0\quad 0)$
$\vee$	$\begin{pmatrix}0&1\\1&1\end{pmatrix}\underline{\begin{pmatrix}0\\1\end{pmatrix}} = \begin{pmatrix}1\\1\end{pmatrix}$	$(0\quad 1)\underline{\begin{pmatrix}0&1\\1&1\end{pmatrix}} = (1\quad 1)$
$\uparrow$	$\begin{pmatrix}1&0\\0&0\end{pmatrix}\underline{\begin{pmatrix}0\\1\end{pmatrix}} = \begin{pmatrix}0\\0\end{pmatrix}$	$(0\quad 1)\underline{\begin{pmatrix}1&0\\0&0\end{pmatrix}} = (0\quad 0)$
$\downarrow$	$\begin{pmatrix}1&1\\1&0\end{pmatrix}\overline{\begin{pmatrix}1\\0\end{pmatrix}} = \begin{pmatrix}1\\1\end{pmatrix}$	$(1\quad 0)\overline{\begin{pmatrix}1&1\\1&0\end{pmatrix}} = (1\quad 1)$
$\rightarrow$	$\begin{pmatrix}1&1\\0&1\end{pmatrix}\underline{\begin{pmatrix}0\\1\end{pmatrix}} = \begin{pmatrix}1\\1\end{pmatrix}$	$(1\quad 0)\overline{\begin{pmatrix}1&1\\0&1\end{pmatrix}} = (1\quad 1)$
$\leftarrow$	$\begin{pmatrix}1&0\\1&1\end{pmatrix}\overline{\begin{pmatrix}1\\0\end{pmatrix}} = \begin{pmatrix}1\\1\end{pmatrix}$	$(0\quad 1)\underline{\begin{pmatrix}1&0\\1&1\end{pmatrix}} = (1\quad 1)$
$\backslash$	$\begin{pmatrix}0&0\\1&0\end{pmatrix}\underline{\begin{pmatrix}0\\1\end{pmatrix}} = \begin{pmatrix}0\\0\end{pmatrix}$	$(1\quad 0)\overline{\begin{pmatrix}0&0\\1&0\end{pmatrix}} = (0\quad 0)$
$/$	$\begin{pmatrix}0&1\\0&0\end{pmatrix}\overline{\begin{pmatrix}1\\0\end{pmatrix}} = \begin{pmatrix}0\\0\end{pmatrix}$	$(0\quad 1)\underline{\begin{pmatrix}0&1\\0&0\end{pmatrix}} = (0\quad 0)$

According to the analysis given above the only logic which operates strictly in terms of true or false, in both left and right logics, and without the emergence of undecided truth-vectors, is the logic of matrix equivalence and matrix negation $\{\leftrightarrow, \rceil\}$.

Multiplication table

In conventional logic compound expressions are formed from alternating logic variables and logic connectives. The only exception is the unary connective negation, which can be preceded by some other connective. All other situations when one connective is followed by another connective have no meaning and are rejected as not being well-formed formulas. Actually, the exceptional case of negation does not imply an interaction of the negation connective with the binary connective which precedes it: the negation first must be applied to a logic variable, and only then can a binary connective act between the result obtained and some other logic variable. In marked contrast, matrix logic introduces an important innovation. Here, a logic operator of negation, if placed immediately after another logic operator, acquires the meaning of matrix multiplication. Given the associativity of the latter, it can either act upon a logic vector or interact with the preceding operator, leading in both cases to the identical final result:

$$L(\rceil \mid x> = (L \rceil) \mid x> .$$

Furthermore, we can now generalize this case for any two logic operators, not necessarily involving the negation operator, and consider that any two operators, being brought together, acquire the meaning of matrix multiplication. The product is another logic operator which is still capable of interacting with another operator or acting upon a logic vector.
Complete information about the interactions of logic operators is contained in the multiplication table of matrix logic given below in table 3.1:

	↔	⌉	∧	∨	→	←	↓	↑	\	/	⊥	⊤	⊣	⊢	1	0
↔	↔	⌉	∧	∨	→	←	↓	↑	\	/	⊥	⊤	⊣	⊢	1	0
⌉	⌉	↔	/	→	∨	↓	←	\	↑	∧	⊤	⊥	⊣	⊢	1	0
∧	∧	\	∧	⊥	∧	⊥	\	0	\	0	⊥	0	∧	\	⊥	0
∨	∨	←	⊣					\	⊢	∧	1	⊥				0
→	→	↓	⊣					↑	⊢	/	1	⊤				0
←	←	∨	∧					⊢	\	⊣	⊥	1				0
↓	↓	→	/					⊢	↑	⊣	⊤	1				0
↑	↑	/	0	/	⊤	↑	⊤	↑	0	/	0	⊤	/	↑	⊤	0
\	\	∧	0	∧	⊥	\	⊥	\	0	∧	0	⊥	∧	\	⊥	0
/	/	↑	/	⊤	/	⊤	↑	0	↑	0	⊤	0	/	↑	⊤	0
⊥	⊥	⊥	∧					\	\	∧	⊥	⊥				0
⊤	⊤	⊤	/					↑	↑	/	⊤	⊤				0
⊣	⊣	⊢	⊣	1	⊣	1	⊢	0	⊢	0	1	0	⊣	⊢	1	0
⊢	⊢	⊣	0	⊣	1	⊢	1	⊢	0	⊣	0	1	⊣	⊢	1	0
1	1	1	⊣					⊢	⊢	⊣	1	1				0
0	0	0	0	0	0	0	0	0	0	0	0	0	0	0	0	0

Table 3.1 Multiplication table of matrix logic

To be consistent with the matrix logic approach, we must exclude all those interactions which violate the condition of binarity of the product. These forbidden interactions are indicated in the multiplication table by empty squares. Imposition of such a restriction implies that interaction of any two logic operators again produces a logic operator.

As a consequence of this constraint we must exclude from the 256 possible interactions 49 interactions as forbidden by matrix logic. The operational set of the latter thus consists of 207 multiplicative interactions.

The multiplicative table clearly shows the special role played by the subset of identity and negation $\{\leftrightarrow, \urcorner\}$: interacting with the set Ω^{+}, identity and negation, uniquely map the latter on itself. Apart from interactions with $\{\leftrightarrow, \urcorner\}$, interactions of the remaining logic operators with the set Ω^{+} reduce the latter to a decuplet subset consisting of the following 10 operators:

$$\{\wedge, /, \backslash, \uparrow, \bot, \top, \dashv, \vdash, \hat{1}, \hat{0}\}$$

A great deal of useful information concerning the behaviour and properties of logic operators can be deduced from the multiplication table of matrix logic, and will be explicitly demonstrated by further material.

Major basis set

We have seen that in conventional logic any logic formula can be expressed by the major basis set of connectives $\{\urcorner, \wedge, \vee\}$. As follows from the multiplication table, the same holds true in matrix logic for the corresponding logic operators. Explicitly, we have the following equalities:

$$\begin{array}{rclcrcl}
\urcorner &=& \urcorner & \qquad & \uparrow &=& \urcorner\wedge\urcorner \\
\wedge &=& \wedge & & \downarrow &=& \urcorner\vee\urcorner \\
\vee &=& \vee & & \bot &=& \wedge\vee \\
\leftrightarrow &=& \urcorner\urcorner & & \dashv &=& \vee\wedge \\
\rightarrow &=& \urcorner\vee & & \vdash &=& \urcorner\vee\wedge\urcorner \\
\leftarrow &=& \vee\urcorner & & \hat{1} &=& \vee\wedge\vee \\
\backslash &=& \wedge\urcorner & & \hat{0} &=& \wedge\urcorner\wedge \\
/ &=& \urcorner\wedge & & \top &=& \urcorner\wedge\vee\urcorner
\end{array}$$

We have generated the complete set of logic operators Ω^{+} from three basic operators:

$$\{\urcorner, \wedge, \vee\}.$$

Complement operator and matrix negation

In the previous section we emphasized the special role of the matrix negation operator $\rceil$. It incorporates in itself the properties of exclusive disjunction, inequivalence, and the operation addition mod2. Viewed from a group-theoretic viewpoint, matrix negation is the only group operation on the set Ω^+. This list of the remarkable properties of matrix negation is not yet complete; it will now be enlarged through a consideration of the complement operator as a generalization of matrix negation.
At first it will be useful to recall that the negation operator $\rceil$, acting upon a logical vector, produces a complementary vector:

$$\rceil |0\rangle = |1\rangle$$

$$\rceil |1\rangle = |0\rangle$$

where $|0\rangle$, $|1\rangle$ abbreviate (1,0) and (0,1) respectively.
Thus, in case of logic vectors, which are either (1×2) or (2×1) matrices, the operator $\rceil$ in fact acts as a complement operator. Seeking the most general definition for the complement operator, we will require that the complement operator must act in a similar manner upon all matrices used by matrix logic: these are (1×1), (1×2), (2×1), and (2×2) matrices. In other words, we are proposing a generalization of negation operation that would incorporate conventional Boolean scalars, which are (1×1) matrices; the logic vectors complementation discussed above; and, finally, complement operation applied to (2×2) L-operators.
This matrix logic complement operation can be formalized in the following relation:

$$\overline{\alpha} = \mathbf{1} - \alpha$$

where α and $\overline{\alpha}$ are the given matrix and its complement respectively, $\mathbf{1}$ – is a corresponding unit constant, which becomes a number 1 if α is a scalar; a vector (1,1) if α is an L-vector; and the matrix $\begin{pmatrix} 1 & 1 \\ 1 & 1 \end{pmatrix}$ if α is an L-operator.

The important relation which unifies the three different aspects of the complement operator can be proved in matrix logic. Namely, a complement of a bilinear product is equal to a bilinear product of the complementary operator:

$$\overline{\langle x | L | y \rangle} = \langle x | \overline{L} | y \rangle.$$

Indeed:

$$1 - \langle x \mid L \mid y\rangle = \langle x \mid \hat{1} - L \mid y\rangle =$$
$$= \langle x \mid \hat{1} \mid y\rangle - \langle x \mid L \mid y\rangle = 1 - \langle x \mid L \mid y\rangle .$$

The complement operator can be carried in and out of the bilinear brackets. By computing the logic expression on the left side of the equation, we obtain a scalar. A negation or complement of this scalar is equal to a bilinear product computed for an operator which is a complement of the given operator.

The important property of the complement operator as a generalization of negation consists in its ability to act in one of three different ways. In dealing with compound logic expressions we can choose the most useful way. For example, in the expression

$$\overline{(x \downarrow y)} \vee (\bar{x} \wedge \bar{y})$$

we can choose to compute the expression $(x \downarrow y)$ and then apply the complement operation to the scalar thus obtained :

$$\overline{(x \downarrow y)} \vee (\bar{x} \wedge \bar{y}) = \bar{0} \vee (\bar{x} \wedge \bar{y}) = 1 \vee (\bar{x} \wedge \bar{y}) .$$

We also can adopt a different route, taking into account the multiplicative property of the complement operator, which allows the latter to act as matrix negation. In such a case we derive the following result:

$$\overline{(x \downarrow y)} \vee (\bar{x} \wedge \bar{y}) = (x \downarrow y) \urcorner \vee (\bar{x} \wedge \bar{y}) = (x \downarrow y) \rightarrow (\bar{x} \wedge \bar{y})$$

And, finally, we can apply the complement operator to a logic operator $\downarrow$, carrying the complement operation inside the brackets:

$$\overline{(x \downarrow y)} \vee (\bar{x} \wedge \bar{y}) = (x \wedge y) \vee (\bar{x} \wedge \bar{y}) .$$

Independently of the way in which the complement operator is applied, the resulting expressions are logically equivalent and must yield the same numerical values.

Complementary subsets

The properties of the complement operator make it a very useful tool for the transformation of matrix logic expressions. It is important to bear in mind not only the similarity but also the difference in the application of the complement operator to a logic scalar, logic vector and logic operator:

$$\bar{x} = 1 - x = \neg x ,$$

$$|\bar{x}\rangle = 1 - |x\rangle = \neg | x\rangle ,$$

$$\bar{L} = \hat{1} - L \neq \neg L .$$

Whereas complementation of a logic vector is equal to a product of a vector with the negation operator, this is not the case with the logic operators. Left and right applications of the negation operator to the set Ω^+ reproduce the set Ω^+.

Left application

{ $\leftrightarrow$	$\neg$ }
{ $\vee$	$\rightarrow$ }
{ $\leftarrow$	$\downarrow$ }
{ $\uparrow$	$\backslash$ }
{ $\wedge$	$/$ }
{ $\perp$	$\top$ }
{ $\dashv$	$\dashv$ }
{ $\vdash$	$\vdash$ }
{ $\hat{1}$	$\hat{1}$ }
{ $\hat{0}$	$\hat{0}$ }

Right application

{ $\leftrightarrow$	$\neg$ }
{ $\vee$	$\leftarrow$ }
{ $\rightarrow$	$\downarrow$ }
{ $\uparrow$	$/$ }
{ $\wedge$	$\backslash$ }
{ $\dashv$	$\vdash$ }
{ $\perp$	$\perp$ }
{ $\top$	$\top$ }
{ $\hat{1}$	$\hat{1}$ }
{ $\hat{0}$	$\hat{0}$ }

On the other hand, application of the complement operation $\bar{L} = \hat{1} - L$ divides the set Ω^+ into two complementary octet subsets

$\{\leftrightarrow\ ,\ \rceil\}$
$\{\wedge\ ,\ \downarrow\}$
$\{\vee\ ,\ \uparrow\}$
$\{\rightarrow\ ,\ \backslash\}$
$\{\leftarrow\ ,\ /\}$
$\{\perp\ ,\ \top\}$
$\{\dashv\ ,\ \vdash\}$
$\{\hat{1}\ ,\ \hat{0}\}$

Application of logic operators to a logic vector is characterized by the fact that the complement of an operator-vector product is equal to the product with the complementary operator:

$$\bar{L}\mid x> = \overline{L\mid x>}.$$

Similarly, in bra vector space:

$$<x\mid\bar{L} = \overline{<x\mid L}\ .$$

Indeed,

$$\bar{L}\mid x> = (\hat{1} - L)\mid x> = \hat{1}\mid x> - L\mid x> = \overline{L\mid x>}\ .$$

Clearly

$$\bar{\bar{L}} = \overline{\hat{1} - L} = L\ .$$

A relation between a pair of complementary logic vectors can be described by a matrix equation:

$$L\mid x> = L\rceil\mid\bar{x}>,$$

and in bra vector space:

$$<x\mid L = <\bar{x}\mid\rceil L$$

where $\{L, L\rceil\}$ and $\{\rceil L, L\}$ are corresponding pairs of logic operators.

In spite of the uniformity of the description, three different outcomes are possible, depending on the result of multiplication $L\rceil$ or $\rceil L$. The set Ω^+ is divided into three subsets according to

$$L\urcorner = \begin{cases} \bar{L} & \text{– complementary product} \\ L & \text{– invariant product} \\ L' & \text{– neither invariant nor complementary.} \end{cases}$$

A similar division obtains in the left application set.

The minor basis sets and orthogonal transformation of logic operators

In conventional logic the major basis set $\{\urcorner, \wedge, \vee\}$ can be reduced to the minor basis sets:

$$\{\urcorner, \wedge\} \text{ or } \{\urcorner, \vee\}.$$

This reduction is possible because of the De Morgan duality rules:

$$\overline{x \wedge y} = \bar{x} \vee \bar{y}$$

$$\overline{x \vee y} = \bar{x} \wedge \bar{y}$$

Introduction of the complement operator, as proposed in the previous section, makes it possible to formulate a matrix equivalent of the De Morgan rules.

The matrix formulation reveals new properties of these rules and leads to a formulation of the generalized De Morgan rule.

To substantiate these claims we need to analyse yet another fundamental property of matrix negation. In the set Ω^+ the negation operator $\urcorner$, together with the trivial case of the identity operator, is the only operator of orthogonal transformation . As we recall from chapter 1, for such operators the transpose is equal to the inverse. This is, indeed, the case with the negation operator:

$$\tilde{\urcorner} = \urcorner^{-1} = \urcorner.$$

Taking into account this fact and the definition of the complement operator, it is possible to show that the duality De Morgan rules of conventional logic take in matrix logic the form of the anti-orthogonal transformation, as the following derivations demonstrate:

$$\overline{\langle x \mid \wedge \mid y \rangle} = \langle x \mid \downarrow \mid y \rangle = \langle x \mid \urcorner \vee \urcorner \mid y \rangle = \langle \bar{x} \mid \vee \mid \bar{y} \rangle$$

$$\overline{\langle x \mid \vee \mid y \rangle} = \langle x \mid \uparrow \mid y \rangle = \langle x \mid \urcorner \wedge \urcorner \mid y \rangle = \langle \bar{x} \mid \wedge \mid \bar{y} \rangle$$

This proves that reduction of the major basis set to one of the minor basis sets in matrix logic is achieved by anti-orthogonal transformations.

Clearly, the anti-orthogonal transformation can be considered not only for the operators of conjunction and disjunction but for any logic operator in the set Ω^+. In this way we arrive at the generalized duality rule, which in matrix logic takes the uniform guise of anti-orthogonal transformation.

Within the general anti-orthogonal transform of matrix logic three distinct classes can be specified, and their appearance is due to the three different outcomes of the orthogonal transformation θ:

$$\theta\, L = \rceil L \rceil = L \quad \text{– invariant product}$$

$$\theta\, L = \rceil L \rceil = \bar{L} \quad \text{– complementary product}$$

$$\theta\, L = \rceil L \rceil = L' \neq L \neq \bar{L} \quad \text{– not invariant, not complementary}$$

The following chart presents the products of orthogonal and anti-orthogonal transformations in the set Ω^+, which is, naturally, closed under both operations:

L	θL	$\overline{\theta L}$
$\leftrightarrow$	$\leftrightarrow$	$\rceil$
$\rceil$	$\rceil$	$\leftrightarrow$
$\wedge$	$\uparrow$	$\vee$
$\vee$	$\downarrow$	$\wedge$
$\rightarrow$	$\leftarrow$	$/$
$\leftarrow$	$\rightarrow$	$\backslash$
$\backslash$	$/$	$\leftarrow$
$/$	$\backslash$	$\rightarrow$
$\uparrow$	$\wedge$	$\downarrow$
$\downarrow$	$\vee$	$\uparrow$
$\perp$	$\top$	$\perp$
$\dashv$	$\vdash$	$\dashv$
$\vdash$	$\dashv$	$\vdash$
$\hat{1}$	$\hat{1}$	$\hat{0}$
$\hat{0}$	$\hat{0}$	$\hat{1}$

Table 3.2 Orthogonal and anti-orthogonal transforms in matrix logic. Anti-orthogonal transforms of L-operators correspond to the dual connectives in conventional logic.

The anti-orthogonal transformation of logic operators is a formalization of the empirical notion of dual connectives in conventional logic.

The three different responses of logic operators to orthogonal transformation allow us to derive three subsets of duality rules:

<u>Subset $\overline{\theta L} = \overline{L}$</u>

1. $\overline{\langle x | \leftrightarrow | y \rangle} = \langle x | \urcorner | y \rangle = \langle \bar{x} | \urcorner | \bar{y} \rangle$
2. $\overline{\langle x | \urcorner | y \rangle} = \langle x | \leftrightarrow | y \rangle = \langle \bar{x} | \leftrightarrow | \bar{y} \rangle$
3. $\overline{\langle x | \hat{1} | y \rangle} = \langle x | \hat{0} | y \rangle = \langle \bar{x} | \hat{0} | \bar{y} \rangle$
4. $\overline{\langle x | \hat{0} | y \rangle} = \langle x | \hat{1} | y \rangle = \langle \bar{x} | \hat{1} | \bar{y} \rangle$

<u>Subset $\overline{\theta L} = L$</u>

1. $\overline{\langle x | \perp | y \rangle} = \langle x | \top | y \rangle = \langle \bar{x} | \perp | \bar{y} \rangle$
2. $\overline{\langle x | \top | y \rangle} = \langle x | \perp | y \rangle = \langle \bar{x} | \top | \bar{y} \rangle$
3. $\overline{\langle x | \dashv | y \rangle} = \langle x | \vdash | y \rangle = \langle \bar{x} | \dashv | \bar{y} \rangle$
4. $\overline{\langle x | \vdash | y \rangle} = \langle x | \dashv | y \rangle = \langle \bar{x} | \vdash | \bar{y} \rangle$

<u>Subset $\overline{\theta L} = L' \neq \overline{L} \neq L$</u>

1. $\overline{\langle x | \wedge | y \rangle} = \langle x | \downarrow | y \rangle = \langle \bar{x} | \vee | \bar{y} \rangle$
2. $\overline{\langle x | \vee | y \rangle} = \langle x | \uparrow | y \rangle = \langle \bar{x} | \wedge | \bar{y} \rangle$
3. $\overline{\langle x | \rightarrow | y \rangle} = \langle x | \backslash | y \rangle = \langle \bar{x} | / | \bar{y} \rangle$
4. $\overline{\langle x | \leftarrow | y \rangle} = \langle x | / | y \rangle = \langle \bar{x} | \backslash | \bar{y} \rangle$
5. $\overline{\langle x | \backslash | y \rangle} = \langle x | \rightarrow | y \rangle = \langle \bar{x} | \leftarrow | \bar{y} \rangle$
6. $\overline{\langle x | / | y \rangle} = \langle x | \leftarrow | y \rangle = \langle \bar{x} | \rightarrow | \bar{y} \rangle$
7. $\overline{\langle x | \uparrow | y \rangle} = \langle x | \vee | y \rangle = \langle \bar{x} | \downarrow | \bar{y} \rangle$
8. $\overline{\langle x | \downarrow | y \rangle} = \langle x | \wedge | y \rangle = \langle \bar{x} | \uparrow | \bar{y} \rangle$

Eliminating redundancy, we arrive at the irreducible octet of the anti-orthogonal transformations of matrix logic:

$$\overline{\langle x | \leftrightarrow | y \rangle} = \langle \bar{x} | \urcorner | \bar{y} \rangle$$

$$\overline{\langle x | \hat{1} | y \rangle} = \langle \bar{x} | \hat{0} | \bar{y} \rangle$$

$$\overline{\langle x | \perp | y \rangle} = \langle \bar{x} | \perp | \bar{y} \rangle$$

$$\overline{\langle x | \dashv | y \rangle} = \langle \bar{x} | \dashv | \bar{y} \rangle$$

$$\overline{\langle x | \wedge | y \rangle} = \langle \bar{x} | \vee | \bar{y} \rangle$$

$$\overline{\langle x | \vee | y \rangle} = \langle \bar{x} | \wedge | \bar{y} \rangle$$

$$\overline{\langle x | \rightarrow | y \rangle} = \langle \bar{x} | / | \bar{y} \rangle$$

$$\overline{\langle x | \leftarrow | y \rangle} = \langle \bar{x} | \backslash | \bar{y} \rangle .$$

The material in this section is a good illustration of the new deductive power of matrix logic. Incorporating classical logic as a special case of scalar relaxation and being in full agreement with its results, matrix logic at the same time delivers new results and generalizations. As a formalism with direct computational power, it allows us to uncover those aspects of logic which remain undetected and probably cannot be detected within the framework of conventional logic.

In matrix logic the logic expressions, such as the De Morgan rules, can be viewed as operator equations; like any equation they have a left side and a right side connected by the equality sign. It is important from a computational point of view that this equality denotes genuine algebraic equality and not only an equivalence of truth-values, as would be the case in conventional logic.

As soon as an operator equation is written down, we do not need to bother with what it actually represents in a framework of logical interpretation. For us it is simply a product of the multiplication of three matrices which is equal to a product of the multiplication of another three matrices. As soon as an operator equation is correctly formed we can deal with it on a purely formal mathematical basis.

The significance of this fact reveals itself in the insight that the transformtion rules of conventional logic, the identification of which had somewhat the character of a discovery so

that they even were assigned the name of the discoverer, like the De Morgan rules, the Switcheroo rule, the Shiefer stroke, etc., now can be deduced instantly on purely formal grounds and without any logical evaluation whatsoever.
Furthermore, the mathematically correct transformation of operator equations will produce the complete set of possible logic relations of (2×2) logic operators, including those which are not common in conventional logic.
The question of the transformation of matrix logic equations brings us to the question of transformation rules and inference in matrix logic.

Tranformation rules

The matrix reformulation of logic has far-reaching implications for the ways in which logical inference and the tautology transformation are achieved in logic.

By converting logical relations into matrix operator equations we are not merely introducing new sophisticated symbolic notations, but devising a formalism in which the whole mechanics of reasoning becomes expressible in the language of matrix operator operations. We are now capable not only of transforming logic formulas by traditional logical means but also of employing the operations of matrix algebra.

In conventional logic, the transformation of logical expressions is based on equivalence transformations in which a logic formula or a part of it is replaced by an equivalent formula. The nature of transformations in conventional logic is basically logical though it takes on a certain analytical quality when logic expressions are handled according to the rules of Boolean algebra. In matrix logic the inference of one set of formulas from another set takes the form of rigorous mathematical derivations.
In addition to the scalar rules of Boolean algebra, matrix logic enables us to introduce new and more general tranformation rules which can be characterized as genuinely analytic. There now follows an analysis of some of these rules.

1. Identity decomposition rule

The equivalence operator, representing the identity operation, can always be inserted at any place into a matrix logic expression without causing any change in the latter. However, if the equivalence operator is first presented as a square product of two negation operators:

$$\leftrightarrow = \neg\neg,$$

and is then inserted between a logic operator and a logic vector, it modifies the logic vector into its complement and the logic operator according to the multiplication table. Below, the identity decomposition rule is applied to some matrix logic formulas:

$$\langle x | \vee | y\rangle = \langle x | \urcorner\urcorner\vee | y\rangle = \langle \bar{x} | \rightarrow | y\rangle$$

$$\langle x | \rightarrow | y\rangle = \langle x | \urcorner\urcorner\rightarrow | y\rangle = \langle \bar{x} | \vee | y\rangle$$

$$\langle x | \downarrow | y\rangle = \langle x | \urcorner\urcorner\downarrow | y\rangle = \langle \bar{x} | \leftarrow | y\rangle$$

Inserting the identity operator at another point, we transform the same formulas into the following:

$$\langle x | \vee | y\rangle = \langle x | \vee\urcorner\urcorner | y\rangle = \langle x | \leftarrow | \bar{y}\rangle$$

$$\langle x | \rightarrow | y\rangle = \langle x | \rightarrow\urcorner\urcorner | y\rangle = \langle x | \downarrow | \bar{y}\rangle$$

$$\langle x | \downarrow | y\rangle = \langle x | \downarrow\urcorner\urcorner | y\rangle = \langle x | \rightarrow | \bar{y}\rangle$$

Through a bilateral application of the identity decomposition rule we can transfer negation from the left front vector to the end logic vector:

$$\langle \bar{x} | \wedge | y\rangle = \langle \bar{x} | \theta^2 \wedge | y\rangle = \langle x | \uparrow | \bar{y}\rangle$$

$$\langle \bar{x} | \backslash | y\rangle = \langle \bar{x} | \theta^2 \backslash | y\rangle = \langle x | / | \bar{y}\rangle$$

$$\langle \bar{x} | / | y\rangle = \langle \bar{x} | \theta^2 / | y\rangle = \langle x | \backslash | \bar{y}\rangle$$

$$\langle \bar{x} | \uparrow | y\rangle = \langle \bar{x} | \theta^2 \uparrow | y\rangle = \langle x | \wedge | \bar{y}\rangle$$

The latter transformation can be described as a square of the orthogonal transformation θ^2.

2. Orthogonal transformation rule and complementation rule

We have become acquainted with these rules of matrix logic in the preceding sections. In general form they are written as follows:

$$\langle x | L | y\rangle = \langle \bar{x} | \theta L | \bar{y}\rangle,$$

$$\langle x | L | y\rangle = \overline{\langle x | \bar{L} | y\rangle}.$$

The orthogonal transformation rule actually can be derived in a purely formal way from the identity decomposition rule. The complementation rule follows directly from the definition of the complement operator of matrix logic. The latter rule allows us to extract the negation operator from an operator and either to carry it above the complete logic expression or to multiply it with the first neighbouring operator.

Combined transformation, involving both orthogonal transformation and complementation rules, automatically leads to the general duality transform rule:

$$\langle x \mid L \mid y \rangle = \langle \bar{x} \mid \theta L \mid \bar{y} \rangle = \overline{\langle \bar{x} \mid \overline{\theta L} \mid \bar{y} \rangle} \ .$$

It is possible to form other combined transformations which lead to constructive results and can be used as inference procedures in matrix logic.

3. Transposition transformation rules

The general transposition rule is given by matrix equation:

$$\langle x \mid \overset{\sqcap}{L} \mid y \rangle = \langle \bar{x} \mid \tilde{L} \mid \bar{y} \rangle \ ,$$

which directly follows from two transformation rules:

1. $\langle x \mid L \mid y \rangle = \langle \bar{y} \mid \overset{\sqcap}{L} \mid \bar{x} \rangle$ – reflection rule
2. $\langle x \mid L \mid y \rangle = \langle y \mid \tilde{L} \mid x \rangle$ – transposition rule

where the first represents the reflection or parity rule, to be discussed in greater detail in the next chapter, and the second is the transposition rule, the application of which will be illustrated here.

We shall explore the transposition transformation rule in combination with the orthogonal transformation rule in order to derive the square of contraposition laws.

Contraposition is an important way of applying the implication operator. We are aware that implication differs from conjunction and disjunction in being an asymmetric operator. Therefore

$$\langle x \mid \wedge \mid y \rangle = \langle y \mid \tilde{\wedge} \mid x \rangle = \langle y \mid \wedge \mid x \rangle$$

$$\langle x \mid \vee \mid y \rangle = \langle y \mid \tilde{\vee} \mid x \rangle = \langle y \mid \vee \mid x \rangle$$

but

$$\langle x | \rightarrow | y\rangle = \langle y | \leftarrow | x\rangle \neq \langle y | \rightarrow | x\rangle$$

$$\langle x | \leftarrow | y\rangle = \langle y | \rightarrow | x\rangle \neq \langle x | \leftarrow | y\rangle\ .$$

The converse of implication is different from implication itself, which is the cause of many mistakes ascribable to a failure to distinguish between an implicative expression and its converse. In matrix formulation these are written as

Implication	Converse implication
$\langle x | \rightarrow | y\rangle$	$\langle y | \rightarrow | x\rangle$

To obtain representations for contrapositions from these two expressions, which are converses of each other, we first apply the transposition inference rule:

$$\langle x | \rightarrow | y\rangle = \langle y | \tilde{\rightarrow} | x\rangle = \langle y | \leftarrow | x\rangle$$

$$\langle y | \rightarrow | x\rangle = \langle x | \tilde{\rightarrow} | y\rangle = \langle x | \leftarrow | y\rangle$$

and then the orthogonal transformation inference rule, respectively:

$$\langle y | \leftarrow | x\rangle = \langle \bar{y} | \theta \leftarrow | \bar{x}\rangle = \langle \bar{y} | \rightarrow | \bar{x}\rangle$$

$$\langle x | \leftarrow | y\rangle = \langle \bar{x} | \theta \leftarrow | \bar{y}\rangle = \langle \bar{x} | \rightarrow | \bar{y}\rangle$$

Equating the initial brackets with the derived brackets, we obtain the set of contraposition laws:

$$\langle x | \rightarrow | y\rangle = \langle \bar{y} | \rightarrow | \bar{x}\rangle$$

$$\langle y | \rightarrow | x\rangle = \langle \bar{x} | \rightarrow | \bar{y}\rangle$$

where

$\langle x | \rightarrow | y\rangle$ – implication

$\langle \bar{x} | \rightarrow | \bar{y}\rangle$ – converse contraposition

$\langle y | \rightarrow | x\rangle$ – converse implication

$\langle \bar{y} | \rightarrow | \bar{x}\rangle$ – contraposition

Converse contraposition and contraposition are converses of each other, just as with implication and converse implication. These four implications are related diagonally: contraposition is equal to implication whereas converse contraposition is equal to converse implication.

Implication or conditional can take various forms. It may be an onerous task in some instances to recognize whether we are dealing with implication or its converse. Necessary and sufficient conditions in mathematics can serve as an example: a necessary condition is the converse of a sufficient condition, while their joint validity implies equivalence or biconditional. Difficulties in the phenomenological analysis of implication can also be attributed to the asymmetric structure of implication.

At the same time, in matrix logic, we derive the square of contraposition laws in a purely formal manner, using matrix logic inference rules to transform operator equations. Without any logical evaluation, simply performing matrix operations and manipulating with numbers as dictated by the rules of matrix calculus, we arrive at equations which have meaning and interpretations in a complex logical environment.

The transformation rules provide a simple and effective means of dealing with the operator equations of matrix logic.

Tautologies and contradictions in matrix logic

We have developed a novel interpretation of the logical equalities as matrix operator equations, the solutions of which are represented by appropriate logical vectors, with the final results of computation 1 and 0 corresponding to the 'true' and 'false' of conventional logic.

The final result of computation may or may not depend on the logical vectors substituted into an operator equation.

The matrix logic equations which have been considered so far produce different final results under different substitutions of logic vectors, i.e. under different interpretations, applying conventional terminology.

There are, however, two degenerate classes of matrix logic equations which remain invariant under any possible substitution. All equations of the first class are equivalent to the bilinear product of the tautology operator

$$\langle x \mid \hat{1} \mid y \rangle \equiv 1 \,;$$

all equations of the second class are equivalent to the bilinear product of the empty operator

$$\langle x \mid \hat{0} \mid y \rangle \equiv 0 \,.$$

Independently of the interpretation given to <x| and |y>, the bilinear products remain unchanged, generating in the first case 1, and in the second case 0.
The degenerate operator equations which are equivalent to the first class represent the tautologies of matrix logic, and those equivalent to the second class represent the contradictions.
Consider the following basic examples:

$$<x \mid \vee \mid \bar{x}> \equiv <\bar{x} \mid \vee \mid x> \equiv <x \mid \leftarrow \mid x> \equiv <x \mid \rightarrow \mid x> \equiv 1 ;$$

$$<x \mid \wedge \mid \bar{x}> \equiv <\bar{x} \mid \wedge \mid x> \equiv <x \mid \backslash \mid x> \equiv <x \mid / \mid x> \equiv 0 .$$

Tautologies for compound logic expressions can be formulated in matrix logic as well, and in some instances there are general rules for recognizing them.

Additive properties of logic operators

In the previous section we were primarily concerned with multiplicative aspects of matrix logic. However, when we introduced the complement operator

$$\bar{L} = \hat{1} - L$$

in addition to multiplicative operations we indirectly explored an additive operation.
The additive operations have a broader meaning and, as the example with the complement operator shows, play an important role in the matrix formulation of logic.
When considering the additive properties of logic operators, just as was done for forbidden products, we must also introduce the forbidden additions. In this framework we are defining sort of logical linearity, which implies that the sum of two logic operators is again a logical operator:

$$L_i + L_j = L_k ; \qquad L_i, L_j, L_k \in \Omega^+ .$$

The constraint we are obliged to impose upon additions of logic operators immediately reduces the number of possible sums to a modest set of 50 additions, as compared with 207 possible products.
Among these, several types of additive operations are of fundamental importance in matrix logic.
First let us consider the crossing-over group series of the major basis set $\{\rceil, \wedge, \vee\}$:

$$\neg + \wedge = \vee$$
$$T + \backslash = \downarrow$$
$$\leftrightarrow + / = \rightarrow$$
$$\perp + \uparrow = \leftarrow$$

The operators of the major basis set again appear to be linked, now in an additive way. Since the operator of disjunction is the sum of the operators of negation and conjunction, the bilinear product for disjunction can be written as

$$\langle x \mid \vee \mid y\rangle = \langle x \mid \neg + \wedge \mid y\rangle = \langle x \mid \neg \mid y\rangle + \langle x \mid \wedge \mid y\rangle .$$

An additive decomposition of a bilinear logic product into a sum of the bilinear logic products is applicable to any logical operator expressed via a sum of other logic operators. Since the forbidden additions are excluded by definition, and any addition we consider is closed on the set Ω^+, the constituent terms of a decomposition of a bilinear product are either all equal to 0 or, with the exception of one term, equal to 1.

What is probably the most significant result that emerges from the analysis of the additive properties of matrix logic is a possibility for a radically new definition of the basis set of logic operators. Namely, we can define the basis set in a direct algebraic fashion as an algebraic base, that is, a set of linearly independent matrices from which, with appropriately chosen coefficients, any logic operator from the set Ω^+ can be generated in an additive way. Such a matrix logic base is given by the set of linearly independent operators formed by the conjunction crossing-over series:

$$\wedge = \begin{pmatrix} 0 & 0 \\ 0 & 1 \end{pmatrix}, \quad \backslash = \begin{pmatrix} 0 & 0 \\ 1 & 0 \end{pmatrix}, \quad / = \begin{pmatrix} 0 & 1 \\ 0 & 0 \end{pmatrix}, \quad \uparrow = \begin{pmatrix} 1 & 0 \\ 0 & 0 \end{pmatrix}.$$

Defining coefficients on the set $E_2 = \{0,1\}$ we can generate the complete set of logic operators Ω^+. For example, the disjunction operator will be defined in the matrix logic base as linear combination:

$$\vee = 1 \cdot \wedge + 1 \cdot \backslash + 1 \cdot / + 0 \cdot \uparrow$$

and equivalence as:

$$\leftrightarrow = 1 \cdot \wedge + 0 \cdot \backslash + 0 \cdot / + 1 \cdot \uparrow .$$

With two operations of addition and multiplication, both defined, the set Ω^+ can be derived from the following functionally complete logical bases:

$$\{\wedge, \backslash\ /\}$$
$$\{\backslash\ /, \uparrow\}$$
$$\{\backslash\ /\}$$

With the third operation of matrix transposition added to this list, the functionally complete logical bases can be reduced to two unit sets

$$\{\backslash\} \text{ and } \{/\} .$$

With four matrix operations defined: addition, multiplication, transposition and crossing-over transposition, each of the operators of the additive logical base $\{\wedge, \backslash\ /, \uparrow\}$ becomes functionally complete.

Generators of operators

In the previous sections we proposed placing at the foundation of logic a matrix base in the algebraic sense and showed that such an approach allows one to generate the complete set of logic operators Ω^+.

In spite of the significance of this algebraic matrix approach, there exist even more simple generators of logic.

Contrary to the conventional approach, which requires both logic variables and logic connectives as two distinctly different and two necessary ingredients from which a logic calculus can be developed, matrix logic can be generated from logic vectors alone, without connectives or their matrix equivalents being available in advance. All that is needed to generate the logic in all its complexity is a set of two vectors from two adjoint logic vector spaces:

$$|x\rangle \text{ and } \langle y|$$

The inner product $\langle x \mid y\rangle$, as we know, is the Kronecker function

$$\langle x \mid y\rangle = \delta_{xy}$$

which is a scalar, and which takes us down to the scalar Boolean set $E_2 = \{0,1\}$.

But considering the possible outer products of logic vectors we get

$$|1><1| = \begin{pmatrix} 0 & 0 \\ 0 & 1 \end{pmatrix}$$

$$|1><0| = \begin{pmatrix} 0 & 0 \\ 1 & 0 \end{pmatrix}$$

$$|0><1| = \begin{pmatrix} 0 & 1 \\ 0 & 0 \end{pmatrix}$$

$$|0><0| = \begin{pmatrix} 1 & 0 \\ 0 & 0 \end{pmatrix}.$$

With the help of no more than two dual logic vectors we have generated the set $\{\wedge, \setminus, /, \uparrow\}$, which, as we already know, is the algebraic base of logic, from which any logic operator is derivable as a linear combination with appropriate multipliers. Matrix logic not only transforms conventional connectives into genuine mathematical objects, it also introduces a simple and elegant way to derive them as outer products of the two adjoint vector spaces.

Above all, since logic operators and logic vectors are matrices, with the newly defined generators of logic, the division into logic variables and logic connectives becomes even more artificial.

This result has profound consequences for the design of logic and for an understanding of the nature and origins of logic operations. No less important are the implications for computer science. Whereas in conventional computing theory it is common to divide computing elements into operators and operands, the newly defined generators of logic pave the way to a computing theory in which such a division will be irrelevant.

Summarizing the general results introduced in this chapter, we have presented the interrelation between the scalar, vector, and operator aspects of logic in the following scheme:

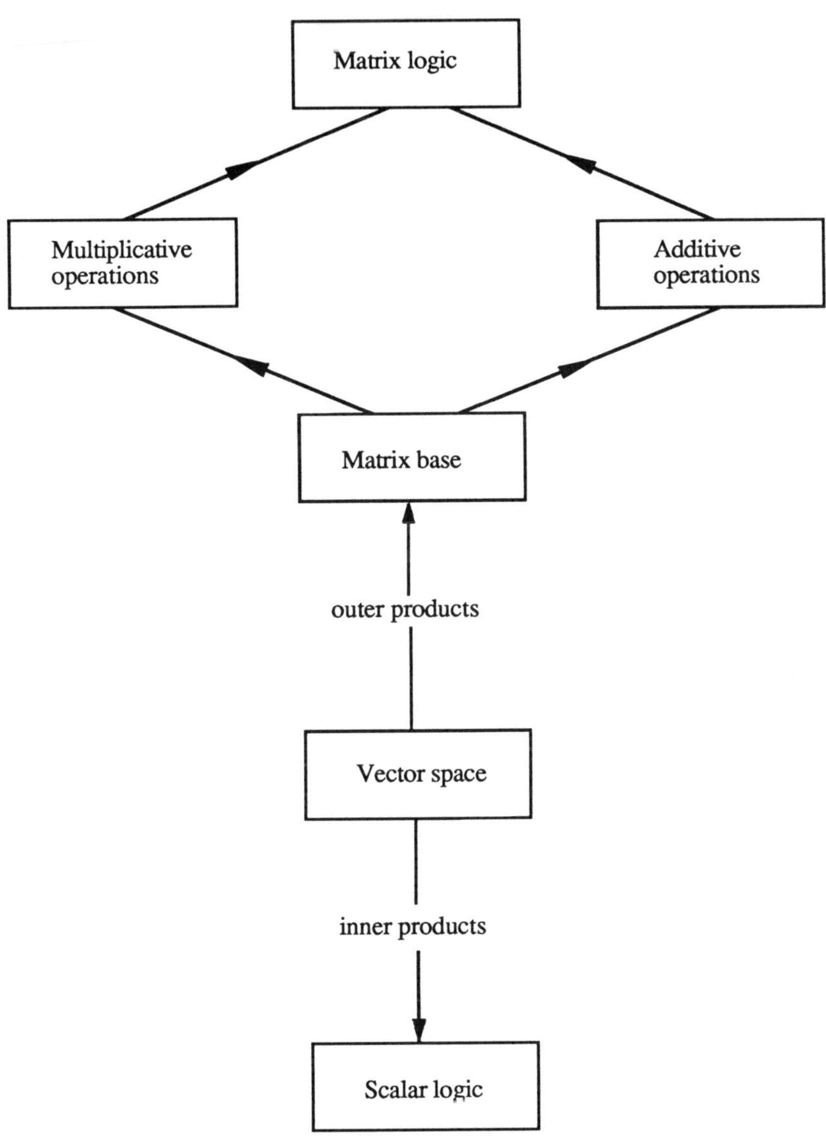

Matrix strings
Compound expressions in matrix logic

The analysis we have made in this chapter shows that matrix logic introduces a new and effective computational method of dealing with the basic logic functions and their interrelations. Our next task is to investigate how matrix logic can be applied to the computation of compound logic expressions or strings. In conventional logic the main method of evaluating large logic strings is based on reducing these to disjunctive or conjunctive normal forms and then minimizing them.

In developing methods for the computation of large logic strings in matrix logic we face not only new opportunities, provided by the richness and power of matrix calculus, but also several major obstacles ascribable to the limitations imposed by the rules of matrix multiplication.

Within the framework of the matrix formulation, a logic string represents the product of a finite number of matrices, that is logic operators and logic vectors, which, to be multiplied, must obey the rules of matrix multiplication. To fulfil the condition of matrix multiplicaton it is necessary that each pair of neighbouring matrices on a string should be compatible, that is, in each succeeding matrix the number of rows must be equal to the number of columns in the preceding matrix. It is not difficult to write down a compound logic expression the corresponding matrix form of which would not comply with the essential condition for matrix multiplication.

In order to overcome this obstacle and make matrix logic strings computable several methods can be formulated.

The following discussion concerns the three central problems of matrix logic, namely, the resolution of 1) operator logic strings; 2) vector-operator strings; and 3) inner product representations.

Operator logic strings

Computation of operator logic strings involves finding a solution to strings of the following structure:

$$\langle x \mid L_1 L_2 L_3 L_4 L_5 L_6 \dots L_n \mid y \rangle$$

where

$$L_1 L_2 L_3 L_4 L_5 L_6 \dots L_n = L_i , \quad i = 1, 16$$

is a product of n logic operators from the set Ω^+, with $n \gg 1$.

yielding either 1 or 0, whereby the string is contracted, bringing together two neighbouring operators, multiplication of which then produces a new operator. In its own turn, the resultant operator, together with the two surrounding logic vectors, forms a nucleus for the next computation. Continuing this process further, we end up with the final result, 1 or 0, and the string vanishes.

As simple as the description of out-from-the-centre computation may look, the actual application of this algorithm requires the resolution of several major issues. These pertain to parity, the forbidden interactions of logic operators, and the associativity of the strings.

Parity. Out-from-the-centre computation is applicable only to properly balanced strings, that is strings in the centre of which we find a logic operator. Naturally this is not generally the case: there exists a set of improperly balanced strings which have at their centre not a logic operator but a logic vector. For the string to be properly balanced it must have an odd number of operators (the negation operator is not taken into account) whereas the number of logic vectors is necessarily even. In an improperly balanced string this relation is reciprocal.

Once again we observe that the question of evenness, or more generally, the question of parity is of great significance in matrix logic. The fact that only properly balanced strings are computable out-from-the-centre in no way prevents us from dealing with improperly balanced strings. However, it is necessary to transform every improper string into a proper string prior to computation. For that purpose we can either expand or contract a string, according to the number of equalities available:

$$x \wedge x = x$$

$$x \vee x = x$$

$$\bar{x} = x \uparrow x = x \downarrow x$$

$$x \wedge (y \vee z) = (x \wedge y) \vee (x \wedge z)$$

$$x \vee (y \wedge z) = (x \vee y) \wedge (x \vee z)$$

$$x \wedge (x \to y) = y\ .$$

Forbidden interactions. Assuming that a properly balanced string is constructed, it may nevertheless not be computable if, at some stage in the contraction of the string, an operator from the bra-branch meets an operator from the ket-branch with which it may not interact (a forbidden interaction). Consequently, to allow computation, no places that are symmetric with respect to the centre of a string may be occupied by operators interaction between which is forbidden.

Consider a properly balanced disjunctive string:

$$\langle a| \vee \langle b| \vee |c\rangle \vee |d\rangle\ .$$

Though properly balanced, the string in its present form is not computable; at the second stage of computation we get

$$\langle a \mid \vee\vee \mid d \rangle \,,$$

which contains the forbidden interaction $\vee^2$; the conjunction operator may not interact with itself.

Nevertheless, computation becomes possible upon transformation by the complementation inference rule:

$$\langle a \mid \vee \langle b \mid \vee \mid c \rangle \vee \mid d \rangle = \overline{\langle \bar{a} \mid \wedge \langle \bar{b} \mid \wedge \mid \bar{c} \rangle \wedge \mid \bar{d} \rangle} \;.$$

<u>Associativity</u>. Associativity is another significant factor to be taken into account when dealing with vector-operator strings. Out-from-the-centre computation is applicable to associative strings where conventional brackets, indicating the order of operations, are unnecessary. For example in the string

$$\langle(\bar{a} \leftarrow b) \vee (c \rightarrow \bar{d})\rangle$$

computation is not possible, not only because at symmetric places there are operators which cannot interact with each other but also because we cannot freely delete the operational brackets due to the non-associativity of the operations involved.

Yet the string can be transformed into a computable one:

$$\langle(\bar{a} \leftarrow b) \vee (c \rightarrow \bar{d})\rangle$$

$$\downarrow$$

$$\langle(a \downarrow b) \vee (c \downarrow d)\rangle$$

$$\downarrow$$

$$\overline{\langle \overline{a \downarrow b} \wedge \overline{c \downarrow d} \rangle}$$

$$\downarrow$$

$$\overline{\langle a \mid \wedge \langle b \mid \wedge \mid c \rangle \wedge \mid d \rangle}$$

$$\downarrow$$

$$\overline{\langle a \mid \wedge \mid d \rangle}$$

$$\downarrow$$

$$1/0$$

We have obtained a computable conjunctive string. Such strings play a central role in the out-from-the-centre computation of vector-operator strings.

It can be shown that in matrix logic any logic expression can be translated into a conjunctive string.

To conclude this section it is worth emphasizing that in spite of its engineering appeal and interesting applications in physics, the role of out-from-the-centre computation in matrix logic is restricted by a large number of limitations.

Inner product representations

An effective computational procedure in matrix logic, applicable both to basic logic functions and to compound logic functions, can be formulated in the form of representations which we shall define as inner product representations.

The simplest case of inner product representations is found for bilinear products of the equivalence operator and the negation operator, where inner product representations are derived directly as follows:

$$\langle x \mid \leftrightarrow \mid y \rangle = \langle x \mid y \rangle$$

$$\langle x \mid \rceil \mid y \rangle = \langle \bar{x} \mid y \rangle = \langle x \mid \bar{y} \rangle = \overline{\langle x \mid y \rangle}\,.$$

We can construct the inner product representations for bilinear products of the operators of the additive matrix base by replacing the operators by corresponding outer products:

$$\langle x \mid \wedge \mid y \rangle = \langle x \mid 1 \rangle\langle 1 \mid y \rangle$$

$$\langle x \mid \backslash \mid y \rangle = \langle x \mid 1 \rangle\langle 0 \mid y \rangle$$

$$\langle x \mid / \mid y \rangle = \langle x \mid 0 \rangle\langle 1 \mid y \rangle$$

$$\langle x \mid \uparrow \mid y \rangle = \langle x \mid 0 \rangle\langle 0 \mid y \rangle$$

In order to derive inner product representations for other operators we either apply the complementation inference rule or express an operator as a product of some other operators:

$$\langle x \mid \downarrow \mid y \rangle = \overline{\langle \bar{x} \mid \uparrow \mid \bar{y} \rangle} = \overline{\langle \bar{x} \mid o \rangle\langle 0 \mid \bar{y} \rangle} = \overline{\langle x \mid 1 \rangle\langle 1 \mid y \rangle}$$

$$\langle x \mid \rightarrow \mid y \rangle = \overline{\langle \bar{x} \mid / \mid \bar{y} \rangle} = \overline{\langle \bar{x} \mid 0 \rangle\langle 1 \mid \bar{y} \rangle} = \overline{\langle x \mid 1 \rangle\langle 0 \mid y \rangle}$$

$$\langle x \mid \leftarrow \mid y \rangle = \overline{\langle \bar{x} \mid \backslash \mid \bar{y} \rangle} = \overline{\langle \bar{x} \mid 1 \rangle\langle 0 \mid \bar{y} \rangle} = \overline{\langle x \mid 0 \rangle\langle 1 \mid y \rangle}$$

$$\langle x \mid \vee \mid y \rangle = \overline{\langle \bar{x} \mid \wedge \mid \bar{y} \rangle} = \overline{\langle \bar{x} \mid 1 \rangle\langle 1 \mid \bar{y} \rangle} = \overline{\langle x \mid 0 \rangle\langle 0 \mid y \rangle}$$

To obtain inner product representations for bilinear products of 1-negation, 0-negation and their converses, we first express these operators as product of operators of the major basis set:

$$<x \mid \perp \mid y> = <x \mid \wedge\vee \mid y> = <x \mid 1>\overline{<0 \mid 1><0 \mid y>} \equiv <x \mid 1>$$

$$<x \mid \top \mid y> = <\bar{x} \mid \wedge\vee \mid \bar{y}> = <x \mid 0>\overline{<0 \mid 1><1 \mid y>} \equiv <x \mid 0>$$

$$<x \mid \dashv \mid y> = <x \mid \vee\wedge \mid y> = \overline{<x \mid 0><1 \mid 0>}<1 \mid y> \equiv <1 \mid y>$$

$$<x \mid \vdash \mid y> = <\bar{x} \mid \vee\wedge \mid \bar{y}> = \overline{<x \mid 1><1 \mid 0>}<0 \mid y> \equiv <0 \mid y>$$

Just like the inner product representations of the equivalence and negation operators, the inner product representations of $\perp$, $\top$, $\dashv$, $\vdash$ are also each reduced to a single inner product.

On the other hand, three inner products appear in the inner product representation of the empty operator:

$$<x \mid \hat{0} \mid y> = <x \mid \wedge\rceil\wedge \mid y> = <x \mid 1><0 \mid 1><1 \mid y>$$

and four inner products appear in the inner product representation of the bilinear product of the tautology operator:

$$<x \mid \hat{1} \mid y> = <x \mid \vee\wedge\vee \mid y> = \overline{<x \mid 0><1 \mid 0>}\,\overline{<0 \mid 1><0 \mid y>}\,.$$

Different logic functions contain different numbers of inner product multipliers: for the basic logic functions it varies from one to four.

Taking into account the fact that any logic operator can be expressed as a multiplicative interaction of other operators, we actually can derive strings with an arbitrary number of inner products, and vice versa a given string can be uniquely convoluted into one of the basic matrix logic functions.

For example, the inner product representation of the conjunction function:

$$<x \mid \wedge \mid y> = <x \mid 1><1 \mid y>\,,$$

can also be written in equivalent form as

$$<x \mid \wedge \mid y> = <x \mid 1>\overline{<1 \mid 1><0 \mid y>}\ ,$$

which is due to the equality

$$\wedge = \backslash\vee\,.$$

It also can be written with as many inner products as desired:

$$<x \mid \wedge \mid y> = <x \mid 1><1 \mid 1><1 \mid 1><1 \mid 1> \ldots <1 \mid y>$$

which is due to the idempotence of the conjunction operator:

$$\wedge^2 = \wedge .$$

The same holds true for the NOR function:

$$<x \mid \uparrow \mid y> = <x \mid 0><0 \mid 0><0 \mid 0><0 \mid 0> \ldots <0 \mid y> .$$

Somewhat different but basically similar expansions of the inner product representation are possible for 1-negation, 0-negation and their converses:

$$<x \mid \dashv \mid y> = <x \mid \dashv^n \mid y> = <1 \mid y>^n$$
$$<x \mid \vdash \mid y> = <x \mid \vdash^n \mid y> = <0 \mid y>^n .$$
$$<x \mid \perp \mid y> = <x \mid \perp^n \mid y> = <x \mid 1>^n$$
$$<x \mid \top \mid y> = <x \mid \top^n \mid y> = <x \mid 0>^n ,$$

where n is any positive integer.

The great advantage of the computation of matrix logic expressions in the form of the inner product representation consists in the possibility of very economical computation. The result of computation is null, if at least one of the inner products is recognized as being composed of complementary vector multipliers, which are either <1 | 0> or <0 | 1>. The result of computation is then predetermined and actual computation becomes unnecessary.
For any scalar string of any possible length, computation must be carried until the appearance, if any, of the first 0, thus significantly reducing the effort in many instances. For example, for the operator logic strings

$$<x \mid /^n \mid y> = <x \mid ////// \ldots / \mid y> = 0$$
$$<x \mid \backslash^n \mid y> = <x \mid \backslash\backslash\backslash\backslash\backslash\backslash \ldots \backslash \mid y> = 0$$

the results follow immediately and computation is not necessary at all. The same applies to the following set of scalar strings:

$$\langle x | 1\rangle\langle 1 | 0\rangle\langle 1 | y\rangle, \qquad \langle x | 1\rangle\langle 0 | 1\rangle\langle 1 | y\rangle,$$
$$\langle x | 1\rangle\langle 1 | 0\rangle\langle 0 | y\rangle, \qquad \langle x | 0\rangle\langle 0 | 1\rangle\langle 1 | y\rangle,$$
$$\langle x | 0\rangle\langle 0 | 1\rangle\langle 0 | y\rangle, \qquad \langle x | 0\rangle\langle 1 | 0\rangle\langle 0 | y\rangle.$$

In order to make our discussion more transparent let us recall the multiplication subtable of the operators of the matrix logic base {∧, \, /, ↑}:

	∧	\	/	↑
∧	∧	\	$\hat{0}$	$\hat{0}$
\	$\hat{0}$	$\hat{0}$	∧	\
/	/	↑	$\hat{0}$	$\hat{0}$
↑	$\hat{0}$	$\hat{0}$	/	↑

We have here an octet of annihilating interactions resulting in the empty operator $\hat{0}$, with the remaining octet reproducing the operators of the matrix base. This may be immediately clear without actual computation, if the base operators are first expressed as outer products, $|1\rangle\langle 1|$, $|1\rangle\langle 0|$, $|0\rangle\langle 1|$, $|0\rangle\langle 0|$ and then attached to each other in various ways. The use of outer products also makes clear how operators of the matrix base respond to transposition:

$$\widetilde{|1\rangle\langle 1|} = |1\rangle\langle 1|$$

$$\widetilde{|1\rangle\langle 0|} = |0\rangle\langle 1|$$

$$\widetilde{|0\rangle\langle 1|} = |1\rangle\langle 0|$$

$$\widetilde{|0\rangle\langle 0|} = |0\rangle\langle 0|,$$

and to orthogonal transformation:

$$\theta\, | 1\rangle\langle 1| = |0\rangle\langle 0|$$
$$\theta\, | 1\rangle\langle 0| = |0\rangle\langle 1| = \widetilde{|1\rangle\langle 0|}$$
$$\theta\, | 0\rangle\langle 1| = |1\rangle\langle 0| = \widetilde{|0\rangle\langle 1|}$$
$$\theta\, | 0\rangle\langle 0| = |1\rangle\langle 1|$$

The strings comprising identical inner products represent the homogeneous scalar strings:

$$\langle x | 1\rangle\langle 1 | 1\rangle\langle 1 | 1\rangle\langle 1 | 1\rangle \ldots \langle 1 | y\rangle = \langle x | \wedge^{n} | y\rangle\,,$$
$$\langle x | 0\rangle\langle 0 | 0\rangle\langle 0 | 0\rangle\langle 0 | 0\rangle \ldots \langle 0 | y\rangle = \langle x | \uparrow^{n} | y\rangle\,.$$

The strings comprising inner products which are complements of each other form the alternating scalar strings:

$$\langle x | \backslash / \backslash / \backslash / \backslash / \ldots \backslash / | y\rangle = \langle x | (\backslash /)^{m} | y\rangle\,,$$
$$\langle x | / \backslash / \backslash / \backslash \ldots / \backslash | y\rangle = \langle x | (\backslash /)^{m} | y\rangle\,.$$

Written explicitly, alternating strings will take the form

$$\langle x | (\backslash /)^{n} | y\rangle = \langle x | 1\rangle\langle 0 | 0\rangle\langle 1 | 1\rangle\langle 0 | 0\rangle \ldots \langle 1 | y\rangle\,,$$
$$\langle z | (/ \backslash)^{n} | y\rangle = \langle x | 0\rangle\langle 1 | 1\rangle\langle 0 | 0\rangle\langle 1 | 1\rangle \ldots \langle 0 | y\rangle\,.$$

Though the pair of homogeneous and the pair of alternating scalar strings differ in their actual fine infrastructure, they nevertheless represent the same logic functions, respectively:

$$\langle x | \wedge^{n} | y\rangle = \langle x | (\backslash /)^{m} | y\rangle$$
$$\langle x | \uparrow^{n} | y\rangle = \langle x | (/ \backslash)^{m} | y\rangle$$

with n, m being any positive integer number, including $n = m$.

Since the above equalities hold for any n and m, the truth-value of homogeneous and alternating strings will be retained if any number of inner products is removed from the strings. Removing outer products reveals the difference between the two kinds of strings. By deleting any outer product we conserve the homogeneous string whereas alternating strings become equal to either $\langle x | / | y\rangle$ or $\langle x | \backslash | y\rangle$ according as the outer product is deleted at the left or at the right end of the string; or alternating strings recombinate each other if the outer products are deleted simultaneously from the left and right ends; or they collapse into an annihilating string, equal to $\langle x | \hat{0} | y\rangle$, if the outer product is deleted from any position inside the string.

An interesting property of the logic strings discussed here is that both a string and its complement yield the same result by computation. Hence, the complementation of a string is the identity transformation, during which the negation operator can be transferred from the front vector to the end vector whereas the string itself undergoes the orthogonal transformation:

for homogeneous strings:

$$<\bar{x} \mid 1><1 \mid 1><1 \mid 1><1 \mid 1><1 \mid 1> \ldots <1 \mid y> =$$

$$<x \mid 0><0 \mid 0><0 \mid 0><0 \mid 0><0 \mid 0> \ldots <0 \mid \bar{y}> .$$

for alternating strings:

$$<\bar{x} \mid 1><0 \mid 1><1 \mid 1><0 \mid 0><1 \mid 1> \ldots <1 \mid y> =$$

$$<x \mid 0><1 \mid 1><0 \mid 0><1 \mid 1><0 \mid 0> \ldots <0 \mid \bar{y}> .$$

These relations have a certain practical interest. Given a dynamic interpretation, the transformations of this type can be viewed as a detachment of negation from the front vector, followed by its propagation through the string until the end vector is reached. The execution of this invariant operation transfers negation to the end logic vector while the string undergoes orthogonal transformation. Simple complementation of the logic strings is thus capable of executing complex logic operations.

Compound logic functions

So far we have studied the inner product representations of the basic logic functions. By using outer product generators in combination with the inference rules of matrix logic we are able to achieve the reduction of any basic function to corresponding inner product representations. However, from the theoretical point of view and the point of view of applications the most important case concerns compound logic functions α containing an arbitrary number of logic variables:

$$\alpha = <x_1 \mid x_2><x_3 \mid x_4><x_5 \mid x_6><x_7 \mid x_8> \ldots$$

where x_1, x_2, x_3 ... are logic vectors or constants, and where some or all inner products can be covered by the complement operator.

For example the conventional logic expressions become in the inner product representation of matrix logic:

$$x \vee (y \rightarrow z) \Rightarrow \overline{\langle x \mid 0\rangle\langle y \mid 1\rangle\langle 0 \mid z\rangle}$$

$$x \wedge (y \rightarrow z \Rightarrow \langle x \mid 1\rangle\overline{\langle y \mid 1\rangle\langle 0 \mid z\rangle}$$

$$\overline{x \rightarrow (y \wedge z)} \Rightarrow \langle x \mid 1\rangle\overline{\langle y \mid 1\rangle\langle 1 \mid z\rangle}$$

$$(x \wedge y) \vee z \Rightarrow \overline{\overline{\langle x \mid 1\rangle\langle 1 \mid y}\rangle\langle 0 \mid z\rangle}$$

$$\overline{(\overline{x \vee y}) \rightarrow z} \Rightarrow \langle x \mid 0\rangle\langle 0 \mid y\rangle\langle 0 \mid z\rangle$$

If no complement operator appears in the inner product representation, as is the case with the last of the above expressions, a matrix string can be described as a Cartesian product of Kronecker delta functions:

$$\langle x_1 \mid y_1\rangle\langle x_2 \mid y_2\rangle\langle x_3 \mid y_3\rangle \dots \langle x_n \mid y_n\rangle = \delta_{x_1y_1}\,\delta_{x_2y_2}\,\delta_{x_3y_3} \dots \delta_{x_ny_n} \quad .$$

Since any Kronecker delta can be associated with the inner product representation of either the equivalence operator or the negation operator, a reduction to a string of this kind can be considered as a reduction of logic to the logic of equivalence and negation, operating over the inner product representations of matrix logic.

Inner product representations are characterized by a number of remarkable properties which make inference and valuation procedures in matirx logic quite simple, elegant and effective. Of particular importance is that the false solutions are immediately recognizable: if any of the inner products is equal to 0 or any inner product under the complement operator is equal to 1, the total expression is necessarily false and further valuation becomes irrelevant.

Equivalent transformations over the inner product representation

Derivation of logic equivalences over the inner product representation becomes especially easy and transparent. Making use of the two following complementation rules of matrix logic

$$\overline{\langle x \mid y\rangle} = \langle \bar{x} \mid y\rangle = \langle x \mid \bar{y}\rangle$$

$$\overline{\langle x \mid y\rangle\langle z \mid t\rangle} = \langle x \mid \overline{y\rangle\langle z} \mid t\rangle$$

we are able to derive various logic equivalences on a purely formal basis. Consider for example the derivation of the De Morgan relations:

$$\overline{\langle x \mid \wedge \mid y\rangle} = \overline{\langle x \mid 1\rangle\langle 1 \mid y\rangle} = \overline{\langle \bar{x} \mid 0\rangle\langle 0 \mid \bar{y}\rangle} = \langle \bar{x} \mid \vee \mid \bar{y}\rangle$$

$$\overline{\langle x \mid \vee \mid y\rangle} = \overline{\langle x \mid 0\rangle\langle 0 \mid y\rangle} = \langle \bar{x} \mid 1\rangle\langle 1 \mid \bar{y}\rangle = \langle \bar{x} \mid \wedge \mid \bar{y}\rangle$$

Applying the complementation rules in one order for the first equation, and in reverse order for the second equation, we obtain the matrix logic formulation of the De Morgan laws. The nature of these laws becomes especially transparent when observing their pure inner product representations:

$$\langle x \mid 1\rangle\langle 1 \mid y\rangle = \langle \bar{x} \mid 0\rangle\langle 0 \mid \bar{y}\rangle$$

$$\langle x \mid 0\rangle\langle 0 \mid y\rangle = \langle \bar{x} \mid 1\rangle\langle 1 \mid \bar{y}\rangle$$

Clearly, the principles established here can be used for the derivation of a wide range of logic equivalences.

Idempotence and induction

The idempotence of matrix logic is an important factor which paves the way for the introduction of inductive transformational rules.
For the inner product representations the idempotence implies that for any matrix logic expression α:

$$\alpha^2 = \alpha$$

In other words the value of an expression is unaltered if the latter is expanded by any number of identical inner products. And conversely, if an expression contains identical inner products, which are all either under or outside the complement operator, they all, except for one, can be deleted with no effect on the result of evaluation. Consider as an illustration the following reductions:

$$\langle x \mid \wedge \mid x\rangle = \langle x \mid 1\rangle\langle 1 \mid x\rangle = \langle x \mid 1\rangle$$

$$\langle x \mid \vee \mid x\rangle = \overline{\langle \bar{x} \mid 1\rangle\langle 1 \mid \bar{x}\rangle} = \langle x \mid 1\rangle$$

Since it is always possible to expand a logic expression with copies of any inner product which it contains, and of course with any product of the form

$$\langle x \mid x\rangle\,,\ \langle 1 \mid 1\rangle\,,\ \langle 0 \mid 0\rangle\,,$$

we can bring an inductive element into the process of evaluation. Though this induction may not change the logic function realized by a matrix string, but only the length and the structure of a string, such transformation may reveal new properties of an expression, which were not evident prior to induction. This is particularly relevant when a string is expanded with new logic variables.

In view of the commutativity of the inner product, inductive transformations allow one to rearrange the interaction between logic variables and to introduce interaction between previously non-interacting variables, at the same time leaving intact the value of the logic functions as such. For example:

$$\langle x | 1\rangle\langle y | \wedge | z\rangle = \langle x | 1\rangle\langle y | 1\rangle\langle 1 | z\rangle = \langle x | 1\rangle\langle 1 | y\rangle\langle x | 1\rangle\langle 1 | z\rangle = \\ = \langle x | \wedge | y\rangle\langle x | \wedge | z\rangle.$$

$$\langle x | 1\rangle\langle y | \backslash | z\rangle = \langle x | 1\rangle\langle y | 1\rangle\langle 0 | z\rangle = \langle x | 1\rangle\langle 1 | y\rangle\langle x | 1\rangle\langle 0 | z\rangle = \\ = \langle x | \wedge | y\rangle\langle x | \backslash | z\rangle.$$

$$\langle x | 1\rangle\langle y | / | z\rangle = \langle x | 1\rangle\langle y | 0\rangle\langle 1 | z\rangle = \langle x | 1\rangle\langle 0 | y\rangle\langle x | 1\rangle\langle 1 | z\rangle = \\ = \langle x | \backslash | y\rangle\langle x | \wedge | z\rangle.$$

$$\langle x | 1\rangle\langle y | \uparrow | z\rangle = \langle x | 1\rangle\langle y | 0\rangle\langle 0 | z\rangle = \langle x | 1\rangle\langle 0 | y\rangle\langle x | 1\rangle\langle 0 | z\rangle = \\ = \langle x | \backslash | y\rangle\langle x | \backslash | z\rangle .$$

It is evident from the derivations we just carried out that the associativity of matrix multiplication and the commutativity of the inner product lead to a sort of logic associativity: a logic vector can be bound via an operator either to a bra logic vector on the left or to a ket logic vector on the right, with both cases yielding the identical result for the evaluation of an expression. For example:

$$\langle x | 1\rangle\langle 0 | y\rangle\langle 1 | z\rangle = \langle x | 1\rangle\langle y | 0\rangle\langle 1 | z\rangle$$

or

$$\langle x | \backslash | y\rangle\langle 1 | z\rangle = \langle x | 1\rangle\langle y | / | z\rangle .$$

Furthermore, inner product representations are characterized not only by commutativity and associativity but also by what may be called superassociativity. An inner product is freely transferable within a logic expression, unless it is prevented by the complement operator from entering or leaving a particular area of a matrix string. Within this constraint superassociativity allows a given inner product or its copy to be detached from a given place on the matrix string and to be attached in a new place, which induces a new interaction. Superassociativity, together with the commutativity of inner products, constitutes in matrix logic a new inference procedure of major importance.

Reducing a given logic expression to the inner product representation, and applying the commutativity, associativity and superassociativity transformations, we can recompose the expression in a great variety of equivalent forms.
This procedure, consisting of reduction followed by multiple convolution, with new variables being added if necessary, provides the foundation for deductive and inductive logic operations.

Variable logic operators

The superassociativity and commutativity of inner products allow us to enrich the notion of logic operators by considering not only constant but also variable logic operators. Both constant and variable logic operators can be generated as the outer products of logic vectors; however, while in the first case the values of the logic vectors are known and constant, in the second case the logic vectors are unknown variables.
Consider the bilinear product of the conjunction logic operator

$$\langle x \mid \wedge \mid y \rangle$$

where $\langle x|$ and $|y\rangle$ are variable logic vectors.
Expressing the conjunction operator via the outer product of two constant logic vectors, we can reduce the bilinear product to the inner product representation

$$\langle x \mid 1\rangle\langle 1 \mid y\rangle \, .$$

In view of the commutativity of the inner product we can write:

$$\langle x \mid 1\rangle\langle 1 \mid y\rangle = \langle 1 \mid x\rangle\langle 1 \mid y\rangle = \langle x \mid 1\rangle\langle y \mid 1\rangle = \langle 1 \mid x\rangle\langle y \mid 1\rangle,$$

where $|x\rangle\langle 1|$, $|1\rangle\langle y|$, $|x\rangle\langle y|$ are variable logic operators. Depending on the values obtained by the vectors $|x\rangle$ and $\langle y|$, the variable logic operator will yield different logic operators. We distinguish atomic and compound variable logic operators. The variable logic operators discussed above are examples of atomic operators. A compound operator is a compound expression of many variables, with one distinguishing feature: it must begin with a ket logic vector and end with a bra logic vector. Compound operators can be found in any compound logic expression.

Reduction theorem

Any logic expression can be reduced to a fundamental product

$$\bigwedge_{j}^{n} \langle x_i | L_j | x_k \rangle \qquad i = 1, m; \quad k = 1, r.$$

where $\langle x_i|$ are vectors or constants from the bra logic vector space, $|x_k\rangle$ are vectors or constants from the ket logic vector space, and L_j are variable or constant logic operators.

A fundamental product is a product of bilinear products where the variable logic vectors $\langle x_i|$, $|x_k\rangle$ and logic operators are not necessarily all different.

The proof of the reduction theorem immediately follows from the convolution of the inner product representation, where the use of variable logic operators may become unavoidable. In performing the convolution of an inner product representation we replace the outer products with the corresponding logic operator, and carry out complement operations if any are present. If the complement operator acts over a single inner product or over a product of two inner products, we know simple rules which allow us to eliminate the complement operator. However, if three or more inner products appear under the complement operator, the situation becomes more complex and may lead to the appearance of compound variable logic operators. In some cases the overall complement can be eliminated due to particular properties of the logic expression, and then the corresponding fundamental product will not contain compound logic operators. The following derivations illustrate two such cases:

$$1.\ \langle\overline{\overline{x | 1\rangle\langle 0 | y}\rangle\langle\overline{x | 0\rangle\langle 1 | y}}\rangle = \langle\overline{x | 1\rangle\langle 1 | y}\rangle\langle\overline{x | 0\rangle\langle 0 | y}\rangle =$$

$$= \langle x | \downarrow | y\rangle\langle x | \vee | y\rangle = \langle\overline{x | y}\rangle$$

$$2.\ \langle\overline{\overline{x | 0\rangle\langle 0 | y}\rangle\langle\overline{x | 1\rangle\langle 1 | y}}\rangle = \langle\overline{x | 1\rangle\langle 0 | y}\rangle\langle\overline{x | 0\rangle\langle 1 | y}\rangle =$$

$$= \langle x | \rightarrow y\rangle\langle x | \leftarrow y\rangle = \langle x | y\rangle$$

The reduction theorem can be considered a central result of matrix logic, similar to reductions to normal forms in conventional logic. Though the fundamental product is a representation that is closely linked with the inner product representation, it has more profound significance, implying that any logic expression can be reduced to a product of the basic valuation devices of matrix logic. In the more general case a fundamental product can be broken down into a sum of fundamental products neither of which involves overall complementation nor compound logic operators. This, however, requires extension of logic into the negative domain, which we are not yet ready to discuss.

Tautologies and contradictions

Two degenerate fundamental products figure prominently in matrix logic. These are the fundamental product of tautology:

$$\overline{\bigwedge_{j}^{n} \langle x_i | L_j | x_k \rangle \; \overline{\bigwedge_{j} \langle x_i | L_j | x_k \rangle}}$$

and the fundamental product of contradiction:

$$\bigwedge_{j}^{n} \langle x_i | L_j | x_k \rangle \; \overline{\bigwedge_{j} \langle x_i | L_j | x_k \rangle}$$

The proof that the above two products are invariant under any interpretation is as follows:

$$\overline{\alpha\overline{\alpha}} = \overline{\alpha(1-\alpha)} = \overline{\alpha-\alpha^{2}} = \overline{\alpha-\alpha} \equiv \overline{0} \equiv 1$$

where α denotes a fundamental product $\bigwedge_{j}^{n} \langle x_i | L_j | x_k \rangle$.

The fourth step in the proof is due to the idempotence property of matrix logic.

Particular reductions of the fundamental products $\overline{\alpha\overline{\alpha}}$ and $\alpha\overline{\alpha}$ give us the whole spectrum of tautologies and contradictions, including those of conventional logic:

Tautologies	Contradictions
$\langle x \mid \leftrightarrow \mid x \rangle = \overline{\langle x \mid \overline{x} \rangle}$	$\langle x \mid \rceil \mid x \rangle = \langle x \mid \overline{x} \rangle$
$\langle x \mid \vee \mid \overline{x} \rangle = \overline{\langle x \mid 1 \rangle \overline{\langle x \mid 1 \rangle}}$	$\langle x \mid \wedge \mid \overline{x} \rangle = \langle x \mid 1 \rangle \overline{\langle x \mid 1 \rangle}$
$\langle x \mid \rightarrow \mid x \rangle = \overline{\langle x \mid 1 \rangle \overline{\langle x \mid 1 \rangle}}$	$\langle x \mid \backslash \mid x \rangle = \langle x \mid 1 \rangle \overline{\langle x \mid 1 \rangle}$

The structure of tautology may not be immediately recognizable in the expression of two logic variables:

$$\overline{\langle x \mid \vee \mid \bar{y}\rangle\langle \bar{x} \mid \vee \mid y\rangle\langle x \mid \vee \mid y\rangle\langle \bar{x} \mid \vee \mid \bar{y}\rangle} \; .$$

However, using the result of the preceding section, we can reduce this expression to an explicit standard form:

$$\overline{\langle x \mid \vee \mid \bar{y}\rangle\langle \bar{x} \mid \vee \mid y\rangle\langle x \mid \vee \mid y\rangle\langle \bar{x} \mid \vee \mid \bar{y}\rangle} = \overline{\langle x \mid y\rangle\overline{\langle x \mid y\rangle}} \; .$$

Reduction of tautology or contradiction to the standard forms may require use of the super-associativity law.

This is the case with the two fundamental inference rules: Modus Ponens and Modus Tollens. In conventional notation these are written:

$$(x \wedge (x \rightarrow y)) \rightarrow y \quad - \quad \text{Modus Ponens}$$

$$((\bar{x} \rightarrow y) \wedge \bar{y}) \rightarrow x \quad - \quad \text{Modus Tollens}$$

In matrix logic formalism they can be expressed respectively as follows

$$\overline{\langle x \mid 1\rangle\langle 0 \mid y\rangle\overline{\langle x \mid 1\rangle\langle 0 \mid y\rangle}}$$

$$\overline{\langle x \mid 0\rangle\langle 0 \mid y\rangle\overline{\langle x \mid 0\rangle\langle 0 \mid y\rangle}} \; ,$$

implying that they belong to the class of tautologies with two logic variables of the general form

$$\overline{\langle x \mid L \mid y\rangle\langle x \mid \bar{L} \mid y\rangle} \; ,$$

where L and $\bar{L}$ are two complementary logic operators.

Orthogonal basis set

At the beginning of this chapter we indicated that the negation operator is of crucial importance in matrix logic. It is the only case in which we gave a new connotation to a classical connective, adding yet another aspect to a complex connective that has already been interpreted as inequivalence, as exclusive disjunction, and as addition mod2. The newly defined operator

of negation, indeed, allowed us to develop a matrix logic formalism consistent with conventional logic. Matrix negation incorporates the properties of the Boolean negation connective. However, in addition it displays new fundamental properties as the operator of the orthogonal transformation, and incorporates some properties of the complement operator.
These universal properties of matrix logic negation were essential in allowing us to achieve reduction of the major basis set $\{\urcorner, \wedge, \vee\}$ to the minor basis sets $\{\urcorner, \wedge\}$ and $\{\urcorner, \vee\}$, and are fundamental for the reduction of matrix logic to its scalar limit - conventional logic.
We now witness yet another important property of matrix negation. Matrix strings in the inner product representation are systems of numerical operations calibrated to respond to two matrix logic operators, equivalence and matrix negation. Because these two operators appear to be sufficient to perform any logic transformation and evaluation, it can be considered that they, together with the system of numerical operations, form a functionally complete set of matrix logic $\{\leftrightarrow, \urcorner\} \cap \{\times, +, -, 0, 1\}$. Since the equivalence operator can be derived as a square product of the negation operator, the set $\{\leftrightarrow, \urcorner\}$ is, actually, redundant and can be reduced to the minimal set $\{\urcorner\} \cap \{\times, +, -, 0, 1\}$ as the fundamental basis set of logic.

The matrix formulation of logic allows us to construct inner product representations which make use of negation as the only logic operation. The role of the fundamental basis set $\{\urcorner\}$ in matrix logic can be compared to the roles of the minimal basis sets $\{\downarrow\}$ and $\{\uparrow\}$ in Boolean logic; however, now the use of a single operation is not made at the expense of clarity and effectiveness, but, on the contrary, leads to vast simplification.

One must of course be aware that we can speak of the set $\{\urcorner\}$ as being functionally complete only in conjunction with the set of numerical operations of multiplication, addition and subtraction: $\{\urcorner\} \cap \{\times,+,-, 0, 1\}$. Later we will show that the subtraction operation can be eliminated from the fundamental basis set.
Another equivalent interpretation is to consider the inner product representation as a mapping of logic defined over the major basis set $\{\urcorner, \wedge, \vee\}$ into a completely numerical system of operations $\{\times, +, -, 0, 1\}$.

Summary of matrix logic algebra

Boolean algebra is designed over the operations of conjunction $\wedge$ and disjunction $\vee$ which are put in correspondence with the operations of multiplication and addition of ordinary algebra. The analysis we have carried out in the preceding sections shows clearly that matrix logic represents more complex relationships between numerical and logical operations.

Although matrix logic algebra is designed similarly over operations of matrix multiplication and addition, these matrix operations are superimposed over logical operations. The domain of matrix logic algebra is incomparably wider and is not restricted to the operations $\wedge$ and $\vee$, but includes the entire set of logic operators. It is important to recognize that whereas in Boolean algebra $\wedge$ and $\vee$ represent operations, in matrix logic algebra these are matrix operators, and there are no particular reasons for their special treatment with respect to other logic operators.
The difference between Boolean and matrix logic algebras can probably best be seen by considering the distributive law. While in Boolean logic algebra, due to the nature of its logic operations, we must distinguish two distributive laws, one for conjunction and one for disjunction

$$x \wedge (y \vee z) = (x \wedge y) \vee (x \wedge z)$$

$$x \vee (y \wedge z) = (x \vee y) \wedge (x \vee z)$$

in matrix logic algebra, due to the nature of matrix operations, two distributive laws

$$\langle x \mid 1 \rangle \langle y \mid \vee \mid z \rangle = \overline{\langle \bar{x} \mid \vee \mid \bar{y} \rangle \langle \bar{x} \mid \vee \mid \bar{z} \rangle}$$

$$\langle \bar{x} \mid 1 \rangle \langle \bar{y} \mid \vee \mid \bar{z} \rangle = \overline{\langle x \mid \vee \mid y \rangle \langle x \mid \vee \mid z \rangle}$$

apparently contain the same information and can be unified into a single equation. On the other hand, in Boolean algebra we consider only the distribution of $\wedge$ over $\vee$ and $\vee$ over $\wedge$. Since in matrix logic the distributive law concerns the matrix operations of multiplication and addition, the distributive law, in the conventional logic sense, can receive multiple realizations. It is possible to consider the "distribution" of some logic operator over some other.
The laws of matrix logic algebra are significally broader than those of Boolean logic, and though the relationships between both are computationally quite simple, conceptually they require a certain effort.
Of primary importance is the fact that matrix logic algebra incorporates Boolean logic algebra as a special scalar limit.
With the convergence of matrix logic to the scalar Boolean domain, matrix operations vanish while matrix operators and logic vectors are replaced by the symbolic notation of conventional logic and two distributive laws emerge in place of one.

The following reiterates in concise form the properties of matrix logic algebra which were discussed in this chapter:

I. Primes

1) $<x \mid y> = <y \mid x>^* = \delta_{xy}$ – logic vector spaces

2) $L_j \in \Omega$ – matrix operators

3) $<x \mid L_j \mid y>$ – valuation device

4) $\{I, \urcorner\} \cap \{\times, +, -, 0, 1\}$ – orthogonal basis set

$\urcorner^{-1} = \tilde{\urcorner}$; $\urcorner^2 = I$

5) $\begin{cases} \overline{\alpha\overline{\alpha}} \text{ – tautology} \\ \alpha\overline{\alpha} \text{ – contradiction} \end{cases}$

II. Inner products

$$<x \mid x> = <\bar{x} \mid \bar{x}> = \overline{\overline{<x \mid x>}} = 1$$

$$\overline{<x \mid x>} = <\bar{x} \mid x> = <x \mid \bar{x}> = 0$$

$$<x \mid y> = <y \mid x>$$

$$<x \mid y>^2 = <x \mid y>$$

$$<x \mid 1> = x \,;\, <x \mid 0> = \bar{x}$$

III. Outer products

$$|x><y| = L$$

$$|y><x| = \tilde{L}$$

$$\overline{|x><y|} = \hat{1} - L$$

$$(|x><x|)^2 = |x><x|$$

$$(|x><\bar{x}|)^2 = (|\bar{x}><x|)^2 = \hat{0}$$

IV. Bilinear products

$$<x \mid L \mid y> = <y \mid \tilde{L} \mid x>$$

$$\overline{<x \mid L \mid y>} = <x \mid \bar{L} \mid y>$$

$$<x \mid L \mid y> = <\bar{x} \mid \theta L \mid \bar{y}>$$

$$<x \mid L^* \mid y> = <\bar{x} \mid \bar{L} \mid \bar{y}>$$

$$\overline{<x \mid 1><0 \mid y>} = <\bar{x} \mid \overline{0><1} \mid \bar{y}>$$

$$<x \mid L \mid y> = <\bar{y} \mid \overset{\sqcap}{L} \mid \bar{x}>$$

V. Fundamental products

$$\bigwedge_j^n <x_i \mid L_j \mid x_k> \qquad i = 1,m \quad k = 1,r$$

3.2 Negative logic Ω^-

In conventional logic there exists the notion of an inverse logic operation: the negation of negation is identity:

$$\bar{\bar{x}} = x \,.$$

This postulate of conventional logic implicitly entails that the negation connective is the inverse of itself, that is, being applied iteratively, it cancels the effect of its previous application. Negation is the only inverse operation known, or actually postulated, in conventional logic. In matrix logic this axiomatic definition of the inverse logic operation is replaced by the rigour of mathematical principle: the square of the negation operator is a unit matrix:

$$\rceil^2 = I \,,$$

which unambiguously implies that the negation operator is a matrix inverse of itself:

$$\top^{-1} = \rceil \,,$$

and therefore

$$\rceil \top^{-1} = \top^{-1} \rceil = \rceil\rceil = I \,.$$

Thus the negation operator, on any even number of applications, cancels itself out.

Since the notion of inverse obtains in matrix logic the constructive meaning of an inverse matrix, this suggests the idea of investigating in this framework the other logic operators from the set Ω^+. As we know, a matrix may have an inverse only if its determinant differs from zero. The table below provides the values of the determinants for the set of logic operators together with the inverses of the non-singular operators.

Operator L	det L	L^{-1}
$\leftrightarrow$	1	$\begin{pmatrix} 1 & 0 \\ 0 & 1 \end{pmatrix}$
$\neg$	−1	$\begin{pmatrix} 0 & 1 \\ 1 & 0 \end{pmatrix}$
$\wedge$	0	–
$\vee$	−1	$\begin{pmatrix} -1 & 1 \\ 1 & 0 \end{pmatrix}$
$\rightarrow$	1	$\begin{pmatrix} 1 & -1 \\ 0 & 1 \end{pmatrix}$
$\leftarrow$	1	$\begin{pmatrix} 1 & 0 \\ -1 & 1 \end{pmatrix}$
$\backslash$	0	–
$/$	0	–
$\uparrow$	0	–
$\downarrow$	−1	$\begin{pmatrix} 0 & 1 \\ 1 & -1 \end{pmatrix}$
$\perp$	0	–
$\top$	0	–
$\dashv$	0	–
$\vdash$	0	–
$\hat{1}$	0	–
$\hat{0}$	0	–

Table 3.3 Inverses of logic operators.

On the set of logic operators Ω^+ we can derive inverses for the six non-singular operators. Two of these, equivalence and negation, are inverses of themselves, and, consequently, their inverses belong to the set Ω^+. But the four remaining inverses, those of disjunction, implication, converse implication and NAND

$$\vee^{-1}, \rightarrow^{-1}, \leftarrow^{-1}, \downarrow^{-1}$$

are operators of a new kind, containing the somewhat confusing negative sign for one of the matrix elements.

An attempt to interpret consistently the negative values of the matrix elements of logic inverses leads to a fundamental extension of the domain of logic. In addition to positive matrix logic, which we described in the first part of this chapter, a negative logic, with very interesting properties, can be developed. The arithmetization of logic has proved the key to a number of important results. The development of Boolean algebra, the Gödel numbering, employed in the proof of the Incompleteness Theorem, and many other results of logic may be ascribed to the formalization of logic. Till now, however, all formalizations of logic have employed positive number representations. The design of negative logic, as the following material will show, reveals new and important facets of logic.

In order to distinguish between the set of operators of positive logic Ω^+ and the set of operators of negative logic Ω^- we assign corresponding superscripts. Similarly, L^+ will indicate an operator from the set Ω^+, and L^- an operator from the set Ω^-.

Conceptually, the essence of negative logic can probably be best understood with the support of an analogy with a physical vacuum. The relation between positive and negative logic values resembles the relationship between positive levels of energy and vacuum negative levels of energy.

The analogy with a physical vacuum is, however, not fully adequate. It would be incorrect to perceive negative logic as a simple mirror inversion of positive logic. The relationship between both logics is more complex. Whereas in Ω^+ logic the values of the components of the logic vectors and matrix elements are defined in the binary set

$$E_2 = \{0,1\} ,$$

we shall define these values in Ω^- logic in the ternary balanced set

$$E_3 = \{-1, 0, 1\} .$$

with the special strictly negative subset of 16 operators defined in the pseudo-binary set

$$E_2^{(-)} = \{-1, 0\} .$$

Because we now have to operate on the set E_3 instead of the familiar binary set E_2, we encounter symmetries of a more complex and richer nature than those of positive logic alone.
To conclude this introductory discussion it may be useful to emphasize the general implications which stem from considering negative logic. On the psychological level one may try to associate positive and negative logics with conscious and unconscious operations respectively. Within the framework of physics negative logic can be consistently interpreted as an anti-logic in which true and false are recombined with each other.
Under this interpretation the operations of negative logic may indeed be thought of as a vacuum sort of operations. Our purpose here, however, is not to pursue these interesting hypotheses, but to investigate the mathematical properties of the negative subset of logic.

Multiplication table of negative logic

The negative inverses which we derived in the preceding section are just four negative logic operators from the much larger set Ω^-. With each operator of positive logic L_i^+, $i = 1, 16$, we can associate a corresponding series of negative logic operators, which by modulo are all equal to some operator L_i^+. Exploring all the combinatorial possibilities of the distribution of negative units within the logic operators we arrive at the following series of operators:

	Ω^+ - logic	Ω^- - logic
$\wedge$ $\backslash$ $/$ $\uparrow$	4 operators	4 operators
$\leftrightarrow$ $\urcorner$	2 operators	6 operators
$\perp$ $\top$ $\dashv$ $\vdash$	4 operators	12 operators

	Ω^+ - logic	Ω^- - logic
$\vee$ $\rightarrow$ $\leftarrow$ $\downarrow$	4 operators	28 operators
$\hat{1}$	1 operator	15 operators
$\hat{0}$	1 operator	1 operator
Total	16 operators	66 operators

The actual structure of the 66 operators of negative logic is as follows:

Negative logic base:

$$\begin{pmatrix} -1 & 0 \\ 0 & 0 \end{pmatrix} \begin{pmatrix} 0 & 0 \\ -1 & 0 \end{pmatrix} \begin{pmatrix} 0 & -1 \\ 0 & 0 \end{pmatrix} \begin{pmatrix} 0 & 0 \\ 0 & -1 \end{pmatrix}$$

Affirmators:

$$\begin{pmatrix} -1 & 0 \\ 0 & 1 \end{pmatrix} \begin{pmatrix} 1 & 0 \\ 0 & -1 \end{pmatrix} \begin{pmatrix} -1 & 0 \\ 0 & -1 \end{pmatrix} \begin{pmatrix} 0 & -1 \\ 1 & 0 \end{pmatrix} \begin{pmatrix} 0 & 1 \\ -1 & 0 \end{pmatrix} \begin{pmatrix} 0 & -1 \\ -1 & 0 \end{pmatrix}$$

Oscillators:

$$\begin{pmatrix} 0 & 0 \\ -1 & 1 \end{pmatrix} \begin{pmatrix} 0 & 0 \\ 1 & -1 \end{pmatrix} \begin{pmatrix} 0 & 0 \\ -1 & -1 \end{pmatrix} \begin{pmatrix} -1 & 1 \\ 0 & 0 \end{pmatrix} \begin{pmatrix} 1 & -1 \\ 0 & 0 \end{pmatrix} \begin{pmatrix} -1 & -1 \\ 0 & 0 \end{pmatrix}$$

$$\begin{pmatrix} 0 & -1 \\ 0 & 1 \end{pmatrix} \begin{pmatrix} 0 & 1 \\ 0 & -1 \end{pmatrix} \begin{pmatrix} 0 & -1 \\ 0 & -1 \end{pmatrix} \begin{pmatrix} -1 & 0 \\ 1 & 0 \end{pmatrix} \begin{pmatrix} 1 & 0 \\ -1 & 0 \end{pmatrix} \begin{pmatrix} -1 & 0 \\ -1 & 0 \end{pmatrix}$$

Negators:

$$\begin{pmatrix} 0 & 1 \\ 1 & -1 \end{pmatrix} \begin{pmatrix} 0 & 1 \\ -1 & 1 \end{pmatrix} \begin{pmatrix} 0 & -1 \\ 1 & 1 \end{pmatrix} \begin{pmatrix} 0 & 1 \\ -1 & -1 \end{pmatrix} \begin{pmatrix} 0 & -1 \\ 1 & -1 \end{pmatrix} \begin{pmatrix} 0 & -1 \\ -1 & 1 \end{pmatrix} \begin{pmatrix} 0 & -1 \\ -1 & -1 \end{pmatrix}$$

$$\begin{pmatrix} -1 & 1 \\ 1 & 0 \end{pmatrix} \begin{pmatrix} 1 & -1 \\ 1 & 0 \end{pmatrix} \begin{pmatrix} 1 & 1 \\ -1 & 0 \end{pmatrix} \begin{pmatrix} -1 & -1 \\ 1 & 0 \end{pmatrix} \begin{pmatrix} -1 & 1 \\ -1 & 0 \end{pmatrix} \begin{pmatrix} 1 & -1 \\ -1 & 0 \end{pmatrix} \begin{pmatrix} -1 & -1 \\ -1 & 0 \end{pmatrix}$$

$$\begin{pmatrix} 1 & 0 \\ -1 & 1 \end{pmatrix} \begin{pmatrix} -1 & 0 \\ 1 & 1 \end{pmatrix} \begin{pmatrix} 1 & 0 \\ 1 & -1 \end{pmatrix} \begin{pmatrix} 1 & 0 \\ -1 & -1 \end{pmatrix} \begin{pmatrix} -1 & 0 \\ -1 & 1 \end{pmatrix} \begin{pmatrix} -1 & 0 \\ 1 & -1 \end{pmatrix} \begin{pmatrix} -1 & 0 \\ -1 & -1 \end{pmatrix}$$

$$\begin{pmatrix} 1 & -1 \\ 0 & 1 \end{pmatrix} \begin{pmatrix} 1 & 1 \\ 0 & -1 \end{pmatrix} \begin{pmatrix} -1 & 1 \\ 0 & 1 \end{pmatrix} \begin{pmatrix} -1 & -1 \\ 0 & 1 \end{pmatrix} \begin{pmatrix} 1 & -1 \\ 0 & -1 \end{pmatrix} \begin{pmatrix} -1 & 1 \\ 0 & -1 \end{pmatrix} \begin{pmatrix} -1 & -1 \\ 0 & -1 \end{pmatrix}$$

Metatautologies:

$$\begin{pmatrix} -1 & 1 \\ 1 & 1 \end{pmatrix} \begin{pmatrix} 1 & 1 \\ -1 & 1 \end{pmatrix} \begin{pmatrix} 1 & -1 \\ 1 & 1 \end{pmatrix} \begin{pmatrix} 1 & 1 \\ 1 & -1 \end{pmatrix}$$

$$\begin{pmatrix} -1 & 1 \\ 1 & -1 \end{pmatrix} \begin{pmatrix} 1 & -1 \\ -1 & 1 \end{pmatrix} \begin{pmatrix} 1 & 1 \\ -1 & -1 \end{pmatrix} \begin{pmatrix} -1 & -1 \\ 1 & -1 \end{pmatrix} \begin{pmatrix} 1 & -1 \\ 1 & -1 \end{pmatrix} \begin{pmatrix} -1 & 1 \\ -1 & 1 \end{pmatrix}$$

$$\begin{pmatrix} 1 & -1 \\ -1 & -1 \end{pmatrix} \begin{pmatrix} -1 & -1 \\ -1 & 1 \end{pmatrix} \begin{pmatrix} -1 & 1 \\ -1 & -1 \end{pmatrix} \begin{pmatrix} -1 & -1 \\ 1 & -1 \end{pmatrix}$$

Negative tautology:

$$\begin{pmatrix} -1 & -1 \\ -1 & -1 \end{pmatrix}$$

Empty operator:

$$\begin{pmatrix} 0 & 0 \\ 0 & 0 \end{pmatrix}$$

The presentation of the multiplication table of negative logic is, for obvious reasons, a significantly larger task then the derivation of the multiplication table for the set Ω^+. By way of illustration let us consider the interaction of the inverses we derived in the previous section with ten operators of positive logic Ω^+:

$\vee^{-1}$	$\downarrow^{-1}$	$\rightarrow^{-1}$	$\leftarrow^{-1}$	
$\vee^{-1}$	$\downarrow^{-1}$	$\rightarrow^{-1}$	$\leftarrow^{-1}$	$\leftrightarrow$
$\rightarrow^{-1}$	$\leftarrow^{-1}$	$\vee^{-1}$	$\downarrow^{-1}$	$\neg$
/	$\begin{pmatrix} 0 & 1 \\ 1 & -1 \end{pmatrix}$	$\begin{pmatrix} 0 & -1 \\ 0 & 1 \end{pmatrix}$	$\wedge$	$\wedge$
$\leftrightarrow$	$\begin{pmatrix} 1 & 1 \\ -1 & 0 \end{pmatrix}$	$\begin{pmatrix} -1 & 0 \\ 1 & 1 \end{pmatrix}$	$\neg$	$\vee$
$\begin{pmatrix} -1 & 0 \\ 1 & 1 \end{pmatrix}$	$\neg$	$\leftrightarrow$	$\begin{pmatrix} 1 & 1 \\ -1 & 0 \end{pmatrix}$	$\rightarrow$
$\neg$	$\begin{pmatrix} 1 & 1 \\ 0 & -1 \end{pmatrix}$	$\begin{pmatrix} 0 & -1 \\ 1 & 1 \end{pmatrix}$	$\leftrightarrow$	$\leftarrow$
$\uparrow$	$\begin{pmatrix} 1 & 0 \\ -1 & 0 \end{pmatrix}$	$\begin{pmatrix} -1 & 0 \\ 1 & 0 \end{pmatrix}$	$\backslash$	$\backslash$
$\begin{pmatrix} 0 & -1 \\ 0 & 1 \end{pmatrix}$	$\wedge$	/	$\begin{pmatrix} 0 & 1 \\ 0 & -1 \end{pmatrix}$	/
$\begin{pmatrix} 0 & -1 \\ 1 & 1 \end{pmatrix}$	$\leftrightarrow$	$\neg$	$\begin{pmatrix} 1 & 1 \\ 0 & -1 \end{pmatrix}$	$\downarrow$

Table 3.4 Left application of inverses

	$\underline{\vee^{-1}}$	$\underline{\downarrow^{-1}}$	$\underline{\rightarrow^{-1}}$	$\underline{\leftarrow^{-1}}$
$\leftrightarrow$	$\underline{\vee^{-1}}$	$\underline{\downarrow^{-1}}$	$\underline{\rightarrow^{-1}}$	$\underline{\leftarrow^{-1}}$
$\urcorner$	$\underline{\leftarrow^{-1}}$	$\underline{\rightarrow^{-1}}$	$\underline{\downarrow^{-1}}$	$\underline{\vee^{-1}}$
$\wedge$	$\backslash$	$\begin{pmatrix} 0 & 0 \\ 1 & -1 \end{pmatrix}$	$\wedge$	$\begin{pmatrix} 0 & 0 \\ -1 & 1 \end{pmatrix}$
$\vee$	$\leftrightarrow$	$\begin{pmatrix} 1 & -1 \\ 1 & 0 \end{pmatrix}$	$\urcorner$	$\begin{pmatrix} -1 & 1 \\ 0 & 1 \end{pmatrix}$
$\rightarrow$	$\urcorner$	$\begin{pmatrix} 1 & 0 \\ 1 & -1 \end{pmatrix}$	$\leftrightarrow$	$\begin{pmatrix} 0 & 1 \\ -1 & 1 \end{pmatrix}$
$\leftarrow$	$\begin{pmatrix} -1 & 1 \\ 0 & 1 \end{pmatrix}$	$\urcorner$	$\begin{pmatrix} 1 & -1 \\ 1 & 0 \end{pmatrix}$	$\leftrightarrow$
$\backslash$	$\begin{pmatrix} 0 & 0 \\ -1 & 1 \end{pmatrix}$	$\wedge$	$\begin{pmatrix} 0 & 0 \\ 1 & -1 \end{pmatrix}$	$\backslash$
$/$	$\uparrow$	$\begin{pmatrix} 1 & -1 \\ 0 & 0 \end{pmatrix}$	$/$	$\begin{pmatrix} -1 & 1 \\ 0 & 0 \end{pmatrix}$
$\uparrow$	$\begin{pmatrix} -1 & 1 \\ 0 & 0 \end{pmatrix}$	$/$	$\begin{pmatrix} 1 & -1 \\ 0 & 0 \end{pmatrix}$	$\uparrow$
$\downarrow$	$\begin{pmatrix} 0 & 1 \\ -1 & 1 \end{pmatrix}$	$\leftrightarrow$	$\begin{pmatrix} 1 & 0 \\ 1 & -1 \end{pmatrix}$	$\urcorner$

Table 3.5 Right application of inverses

According to the multiplication tables the interaction of L^- - and L^+ - operators may result in the appearance of some other L^- -operators or L^+ -operators. In this particular setting we have generated new negative operators; the operators of the matrix logic base; the equivalence operators; and the negation operators.
Since the operator $\rceil$ can now be generated by the interaction of negative and positive operators, its presence in the basis sets of logic ceases to be necessary. For instance, instead of the minor basis sets we can use any of the following sets, which include one of the matrix inverses:

$$\{\overset{-1}{\leftarrow}\ \vee\} \qquad \{\vee\ \overset{-1}{\rightarrow}\}$$

$$\{\overset{-1}{\vee}\ \leftarrow\} \qquad \{\rightarrow\ \overset{-1}{\vee}\}$$

$$\{\overset{-1}{\downarrow}\ \rightarrow\} \qquad \{\leftarrow\ \overset{-1}{\downarrow}\}$$

$$\{\overset{-1}{\rightarrow}\ \downarrow\} \qquad \{\downarrow\ \overset{-1}{\leftarrow}\}$$

In comparison with positive logic, with its 16 operators and multiplication table with 256 squares, negative logic involves 50 operators more, with a multiplication table now consisting of 4356 squares. The complete set of matrix logic, including its positive and negative subsets, contains the empty operator $\hat{0}$ and 80 positive and negative operators, implying that the complete multiplication table of matrix logic has $81 \times 81 = 6561$ squares.

Another important question to consider concerns the interaction of operators of negative logic among themselves. As in positive logic, interaction of L^- - operators also contains a subset of forbidden interactions, which is much greater than in positive logic: 1050 interactions are forbidden. By way of illustration we present in the Table 3.6 the interaction of 15 negative operators from the tautology series. With the exception of 24 empty interactions all the remaining interactions of the tautology series are forbidden.
To obtain the total number of forbidden interactions in the set $\Omega^+ \cap \Omega^-$ we must add to the 1050 forbidden interactions of negative logic, not only the 49 forbidden interactions of positive logic but also another forbidden set which appears as a result of interaction between the operators of positive and negative logics.
Though negative logic contributes significantly to the subset of forbidden interactions the complete logic set is nevertheless incomparably richer in possibilities than the pure positive matrix set.

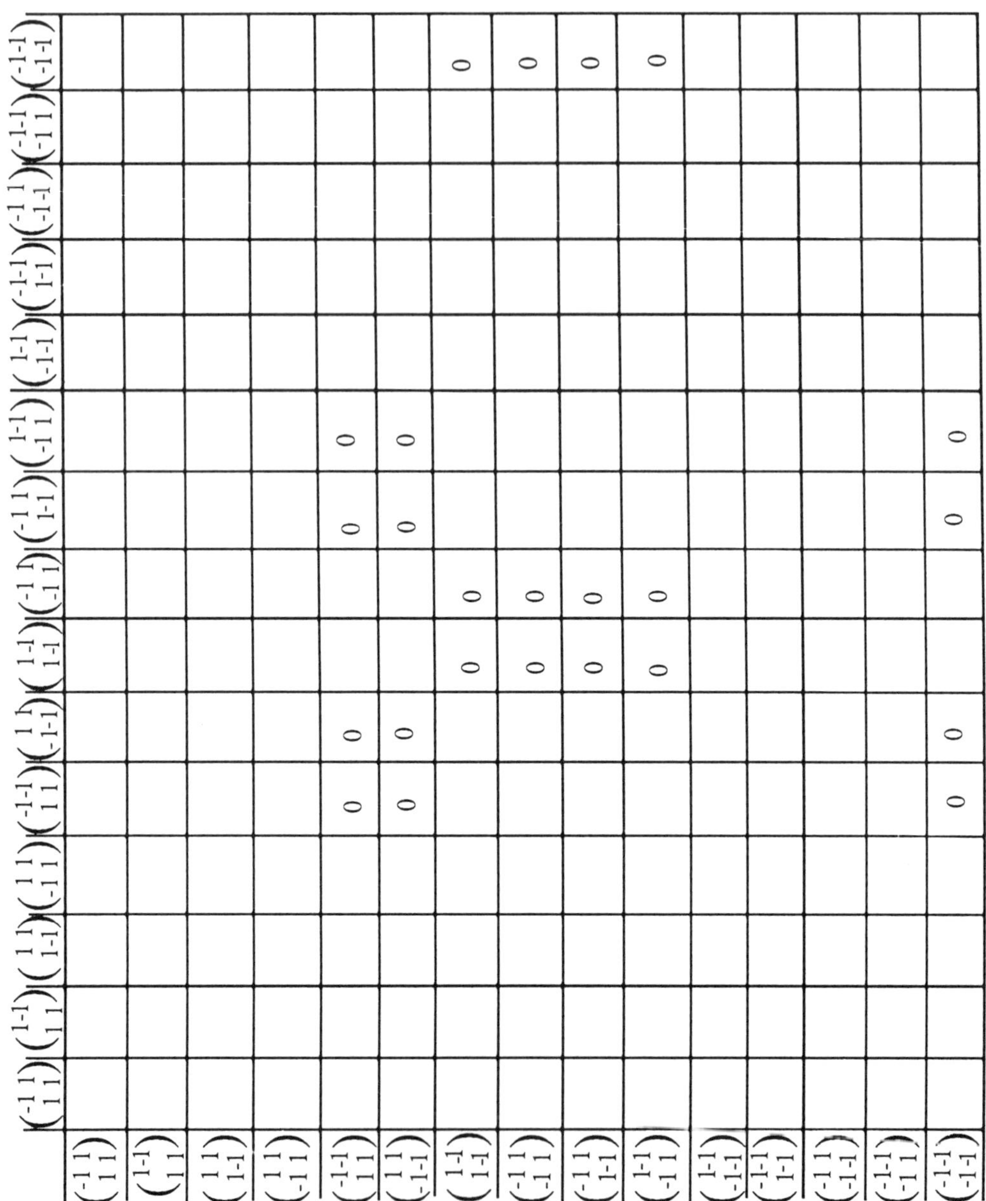

	$\begin{pmatrix}-1&1\\1&1\end{pmatrix}$	$\begin{pmatrix}1&-1\\1&1\end{pmatrix}$	$\begin{pmatrix}1&1\\1&-1\end{pmatrix}$	$\begin{pmatrix}1&1\\-1&1\end{pmatrix}$	$\begin{pmatrix}-1&-1\\1&1\end{pmatrix}$	$\begin{pmatrix}1&1\\-1&-1\end{pmatrix}$	$\begin{pmatrix}1&-1\\1&-1\end{pmatrix}$	$\begin{pmatrix}-1&1\\-1&1\end{pmatrix}$	$\begin{pmatrix}-1&1\\1&-1\end{pmatrix}$	$\begin{pmatrix}1&-1\\-1&1\end{pmatrix}$	$\begin{pmatrix}1&-1\\-1&-1\end{pmatrix}$	$\begin{pmatrix}-1&-1\\1&-1\end{pmatrix}$	$\begin{pmatrix}-1&1\\-1&-1\end{pmatrix}$	$\begin{pmatrix}-1&-1\\-1&1\end{pmatrix}$	$\begin{pmatrix}-1&-1\\-1&-1\end{pmatrix}$
$\begin{pmatrix}-1&1\\1&1\end{pmatrix}$															
$\begin{pmatrix}1&-1\\1&1\end{pmatrix}$															
$\begin{pmatrix}1&1\\1&-1\end{pmatrix}$															
$\begin{pmatrix}1&1\\-1&1\end{pmatrix}$															
$\begin{pmatrix}-1&-1\\1&1\end{pmatrix}$					0	0			0	0					
$\begin{pmatrix}1&1\\-1&-1\end{pmatrix}$					0	0			0	0					
$\begin{pmatrix}1&-1\\1&-1\end{pmatrix}$							0	0							0
$\begin{pmatrix}-1&1\\-1&1\end{pmatrix}$							0	0							0
$\begin{pmatrix}-1&1\\1&-1\end{pmatrix}$							0	0							0
$\begin{pmatrix}1&-1\\-1&1\end{pmatrix}$							0	0							0
$\begin{pmatrix}1&-1\\-1&-1\end{pmatrix}$															
$\begin{pmatrix}-1&-1\\1&-1\end{pmatrix}$															
$\begin{pmatrix}-1&1\\-1&-1\end{pmatrix}$															
$\begin{pmatrix}-1&-1\\-1&1\end{pmatrix}$															
$\begin{pmatrix}-1&-1\\-1&-1\end{pmatrix}$					0	0			0	0					

Table 3.6 Multiplication subtable of the tautology series.

Interaction of negative logic operators among themselves not only generates new negative operators, but is capable of generating positive logic. The possibility of deriving positive logic from the negative operator set makes negative logic in some sense a generalization of positive logic, and offers a way of formally incorporating the latter in the more sophisticated and more complex domain of negative logic values. Most clearly this shows itself in the significant increase of affirmation and negating power in the set $\Omega^{+} \cap \Omega^{-}$.

Affirmation in complete logic

There are just two ways to achieve affirmation in positive logic, by employing either the identity operator or the square of the negation operator.

Derivation of identity as the square product of negation is basically the only non-trivial affirmation in positive logic. With the extension of logic into the negative domain the affirmation power of logic is significantly enlarged: as many as 27 different ways of deriving the identity operator I can now be established:

Positive logic Ω^{+}:

$$I = \begin{pmatrix} 1 & 0 \\ 0 & 1 \end{pmatrix}$$

$$I = \begin{pmatrix} 0 & 1 \\ 1 & 0 \end{pmatrix}^2$$

Negative logic Ω^{-}:

$$I = \begin{pmatrix} -1 & 0 \\ 0 & 1 \end{pmatrix}^2 \qquad I = \begin{pmatrix} 0 & -1 \\ 1 & 0 \end{pmatrix}\begin{pmatrix} 0 & 1 \\ -1 & 0 \end{pmatrix}$$

$$I = \begin{pmatrix} 1 & 0 \\ 0 & -1 \end{pmatrix}^2 \qquad I = \begin{pmatrix} 0 & 1 \\ -1 & 1 \end{pmatrix}\begin{pmatrix} 1 & -1 \\ 1 & 0 \end{pmatrix}$$

$$I = \begin{pmatrix} -1 & 0 \\ 0 & -1 \end{pmatrix}^2 \qquad I = \begin{pmatrix} 0 & -1 \\ 1 & 1 \end{pmatrix}\begin{pmatrix} 1 & 1 \\ -1 & 0 \end{pmatrix}$$

$$I = \begin{pmatrix} 0 & -1 \\ -1 & 0 \end{pmatrix}^2$$

$$I = \begin{pmatrix} -1 & 0 \\ 1 & 1 \end{pmatrix}^2$$

$$I = \begin{pmatrix} 1 & 0 \\ 1 & -1 \end{pmatrix}^2$$

$$I = \begin{pmatrix} 1 & 0 \\ -1 & -1 \end{pmatrix}^2$$

$$I = \begin{pmatrix} -1 & 0 \\ -1 & 1 \end{pmatrix}^2$$

$$I = \begin{pmatrix} 1 & 1 \\ 0 & -1 \end{pmatrix}^2$$

$$I = \begin{pmatrix} -1 & 1 \\ 0 & 1 \end{pmatrix}^2$$

$$I = \begin{pmatrix} -1 & -1 \\ 0 & 1 \end{pmatrix}^2$$

$$I = \begin{pmatrix} 1 & -1 \\ 0 & -1 \end{pmatrix}^2$$

$$I = \begin{pmatrix} 0 & 1 \\ -1 & -1 \end{pmatrix} \begin{pmatrix} -1 & -1 \\ 1 & 0 \end{pmatrix}$$

$$I = \begin{pmatrix} 0 & -1 \\ 1 & -1 \end{pmatrix} \begin{pmatrix} -1 & 1 \\ -1 & 0 \end{pmatrix}$$

$$I = \begin{pmatrix} 0 & -1 \\ -1 & 1 \end{pmatrix} \begin{pmatrix} -1 & -1 \\ -1 & 0 \end{pmatrix}$$

$$I = \begin{pmatrix} 0 & -1 \\ -1 & -1 \end{pmatrix} \begin{pmatrix} 1 & -1 \\ -1 & 0 \end{pmatrix}$$

$$I = \begin{pmatrix} -1 & 0 \\ 1 & -1 \end{pmatrix} \begin{pmatrix} -1 & 0 \\ -1 & -1 \end{pmatrix}$$

$$I = \begin{pmatrix} -1 & 1 \\ 0 & -1 \end{pmatrix} \begin{pmatrix} -1 & -1 \\ 0 & -1 \end{pmatrix}$$

Complete logic $\Omega^{+} \cap \Omega^{-}$:

$$I = \begin{pmatrix} 0 & 1 \\ 1 & 1 \end{pmatrix} \begin{pmatrix} -1 & 1 \\ 1 & 0 \end{pmatrix}$$

$$I = \begin{pmatrix} 1 & 1 \\ 1 & 0 \end{pmatrix} \begin{pmatrix} 0 & 1 \\ 1 & -1 \end{pmatrix}$$

$$I = \begin{pmatrix} 1 & 1 \\ 0 & 1 \end{pmatrix} \begin{pmatrix} 1 & -1 \\ 0 & 1 \end{pmatrix}$$

$$I = \begin{pmatrix} 1 & 0 \\ 1 & 1 \end{pmatrix} \begin{pmatrix} 1 & 0 \\ -1 & 1 \end{pmatrix}$$

The possibility of generating the identity operator not only as the square of the negation operator but in another 25 different new ways allows for a 12.5-fold increase in the affirmation power of logic.

This enrichment of the possibilities of affirmation power is due to the fact that the set of inverses in negative logic is enlarged further. Similar to the subset of positive logic, which has inverses of its own, the subset of negative logic also contains inverses which are pure negative operators. Of these operators, 12 are inverses of themselves and 18 are divided into 9 pairs of inverses. With 4 pairs of cross-talk inverses, that is, the inverses

$$\vee^{-1}, \rightarrow^{-1}, \leftarrow^{-1}, \downarrow^{-1},$$

which act across positive and negative logics, and the 2 inverses of positive logic, altogether in the complete logic set an inverse can be computed for 40 logic operators.

Negating power in complete logic

No product of operators in the set Ω^{+} is capable of generating the negation operator. Because in positive logic negation is a total and non-decomposable operation, we are inclined to think that such is generally the case. However, a simple examination of general logical evaluations shows that apart from all-out complete denials intelligent processes may involve more subtle and incomplete forms of negation. Attempts to deal with this wider concept of negation are made in multivalued logic.

However, these attempts are commonly postulatory in nature and are not derived from rigorously defined principles. The negative extension of logic does offer such a rigorous principle due to the possibility of deriving the negation operator as a product of either two negative operators or negative and positive operators. In addition to the single all-out negation of positive logic in the complete logic set $\Omega^+ \cap \Omega^-$ there are 32 different products generating the negation operator $\rceil$:

<u>24 negative interactions:</u>

$$\begin{pmatrix} 0 & 1 \\ -1 & 1 \end{pmatrix}\begin{pmatrix} -1 & 1 \\ 0 & 1 \end{pmatrix} = \begin{pmatrix} 0 & 1 \\ 1 & 0 \end{pmatrix} \qquad \begin{pmatrix} -1 & 0 \\ 1 & 1 \end{pmatrix}\begin{pmatrix} 0 & -1 \\ 1 & 1 \end{pmatrix} = \begin{pmatrix} 0 & 1 \\ 1 & 0 \end{pmatrix}$$

$$\begin{pmatrix} 0 & -1 \\ 1 & 1 \end{pmatrix}\begin{pmatrix} 1 & 1 \\ 0 & -1 \end{pmatrix} = \begin{pmatrix} 0 & 1 \\ 1 & 0 \end{pmatrix} \qquad \begin{pmatrix} 1 & 0 \\ 1 & -1 \end{pmatrix}\begin{pmatrix} 0 & 1 \\ -1 & 1 \end{pmatrix} = \begin{pmatrix} 0 & 1 \\ 1 & 0 \end{pmatrix}$$

$$\begin{pmatrix} 0 & 1 \\ -1 & -1 \end{pmatrix}\begin{pmatrix} -1 & -1 \\ 0 & 1 \end{pmatrix} = \begin{pmatrix} 0 & 1 \\ 1 & 0 \end{pmatrix} \qquad \begin{pmatrix} -1 & 0 \\ -1 & 1 \end{pmatrix}\begin{pmatrix} 0 & -1 \\ 1 & -1 \end{pmatrix} = \begin{pmatrix} 0 & 1 \\ 1 & 0 \end{pmatrix}$$

$$\begin{pmatrix} 0 & -1 \\ 1 & -1 \end{pmatrix}\begin{pmatrix} 1 & -1 \\ 0 & -1 \end{pmatrix} = \begin{pmatrix} 0 & 1 \\ 1 & 0 \end{pmatrix} \qquad \begin{pmatrix} 1 & 0 \\ -1 & -1 \end{pmatrix}\begin{pmatrix} 0 & 1 \\ -1 & -1 \end{pmatrix} = \begin{pmatrix} 0 & 1 \\ 1 & 0 \end{pmatrix}$$

$$\begin{pmatrix} 0 & -1 \\ -1 & 1 \end{pmatrix}\begin{pmatrix} -1 & -1 \\ 0 & -1 \end{pmatrix} = \begin{pmatrix} 0 & 1 \\ 1 & 0 \end{pmatrix} \qquad \begin{pmatrix} -1 & 0 \\ -1 & -1 \end{pmatrix}\begin{pmatrix} 0 & -1 \\ -1 & 1 \end{pmatrix} = \begin{pmatrix} 0 & 1 \\ 1 & 0 \end{pmatrix}$$

$$\begin{pmatrix} 0 & -1 \\ -1 & -1 \end{pmatrix}\begin{pmatrix} -1 & 1 \\ 0 & -1 \end{pmatrix} = \begin{pmatrix} 0 & 1 \\ 1 & 0 \end{pmatrix} \qquad \begin{pmatrix} -1 & 0 \\ 1 & -1 \end{pmatrix}\begin{pmatrix} 0 & -1 \\ -1 & -1 \end{pmatrix} = \begin{pmatrix} 0 & 1 \\ 1 & 0 \end{pmatrix}$$

$$\begin{pmatrix} 1 & -1 \\ 1 & 0 \end{pmatrix}\begin{pmatrix} 1 & 0 \\ 1 & -1 \end{pmatrix} = \begin{pmatrix} 0 & 1 \\ 1 & 0 \end{pmatrix} \qquad \begin{pmatrix} 1 & 1 \\ 0 & -1 \end{pmatrix}\begin{pmatrix} 1 & 1 \\ -1 & 0 \end{pmatrix} = \begin{pmatrix} 0 & 1 \\ 1 & 0 \end{pmatrix}$$

$$\begin{pmatrix} 1 & 1 \\ -1 & 0 \end{pmatrix}\begin{pmatrix} -1 & 0 \\ 1 & 1 \end{pmatrix} = \begin{pmatrix} 0 & 1 \\ 1 & 0 \end{pmatrix} \qquad \begin{pmatrix} -1 & 1 \\ 0 & 1 \end{pmatrix}\begin{pmatrix} 1 & -1 \\ 1 & 0 \end{pmatrix} = \begin{pmatrix} 0 & 1 \\ 1 & 0 \end{pmatrix}$$

$$\begin{pmatrix} -1 & -1 \\ 1 & 0 \end{pmatrix}\begin{pmatrix} 1 & 0 \\ -1 & -1 \end{pmatrix} = \begin{pmatrix} 0 & 1 \\ 1 & 0 \end{pmatrix} \qquad \begin{pmatrix} 1 & -1 \\ 0 & -1 \end{pmatrix}\begin{pmatrix} -1 & 1 \\ -1 & 0 \end{pmatrix} = \begin{pmatrix} 0 & 1 \\ 1 & 0 \end{pmatrix}$$

$$\begin{pmatrix} -1 & 1 \\ -1 & 0 \end{pmatrix}\begin{pmatrix} -1 & 0 \\ -1 & 1 \end{pmatrix} = \begin{pmatrix} 0 & 1 \\ 1 & 0 \end{pmatrix} \qquad \begin{pmatrix} -1 & -1 \\ 0 & 1 \end{pmatrix}\begin{pmatrix} -1 & -1 \\ 1 & 0 \end{pmatrix} = \begin{pmatrix} 0 & 1 \\ 1 & 0 \end{pmatrix}$$

$$\begin{pmatrix} 1 & -1 \\ -1 & 0 \end{pmatrix}\begin{pmatrix} -1 & 0 \\ -1 & -1 \end{pmatrix} = \begin{pmatrix} 0 & 1 \\ 1 & 0 \end{pmatrix} \qquad \begin{pmatrix} -1 & -1 \\ 0 & -1 \end{pmatrix}\begin{pmatrix} 1 & -1 \\ -1 & 0 \end{pmatrix} = \begin{pmatrix} 0 & 1 \\ 1 & 0 \end{pmatrix}$$

$$\begin{pmatrix} -1 & -1 \\ -1 & 0 \end{pmatrix}\begin{pmatrix} -1 & 0 \\ 1 & -1 \end{pmatrix} = \begin{pmatrix} 0 & 1 \\ 1 & 0 \end{pmatrix} \qquad \begin{pmatrix} -1 & 1 \\ 0 & -1 \end{pmatrix}\begin{pmatrix} -1 & -1 \\ -1 & 0 \end{pmatrix} = \begin{pmatrix} 0 & 1 \\ 1 & 0 \end{pmatrix}$$

8 positive-negative interactions:

$$\begin{pmatrix} 0 & 1 \\ 1 & 1 \end{pmatrix}\begin{pmatrix} 1 & -1 \\ 0 & 1 \end{pmatrix} = \begin{pmatrix} 0 & 1 \\ 1 & 0 \end{pmatrix} \qquad \begin{pmatrix} 1 & 0 \\ -1 & 1 \end{pmatrix}\begin{pmatrix} 0 & 1 \\ 1 & 1 \end{pmatrix} = \begin{pmatrix} 0 & 1 \\ 1 & 0 \end{pmatrix}$$

$$\begin{pmatrix} 1 & 1 \\ 0 & 1 \end{pmatrix}\begin{pmatrix} -1 & 1 \\ 1 & 0 \end{pmatrix} = \begin{pmatrix} 0 & 1 \\ 1 & 0 \end{pmatrix} \qquad \begin{pmatrix} -1 & 1 \\ 1 & 0 \end{pmatrix}\begin{pmatrix} 1 & 0 \\ 1 & 1 \end{pmatrix} = \begin{pmatrix} 0 & 1 \\ 1 & 0 \end{pmatrix}$$

$$\begin{pmatrix} 1 & 0 \\ 1 & 1 \end{pmatrix}\begin{pmatrix} 0 & 1 \\ 1 & -1 \end{pmatrix} = \begin{pmatrix} 0 & 1 \\ 1 & 0 \end{pmatrix} \qquad \begin{pmatrix} 0 & 1 \\ 1 & -1 \end{pmatrix}\begin{pmatrix} 1 & 1 \\ 0 & 1 \end{pmatrix} = \begin{pmatrix} 0 & 1 \\ 1 & 0 \end{pmatrix}$$

$$\begin{pmatrix} 1 & 1 \\ 1 & 0 \end{pmatrix}\begin{pmatrix} 1 & 0 \\ -1 & 1 \end{pmatrix} = \begin{pmatrix} 0 & 1 \\ 1 & 0 \end{pmatrix} \qquad \begin{pmatrix} 1 & -1 \\ 0 & 1 \end{pmatrix}\begin{pmatrix} 1 & 1 \\ 1 & 0 \end{pmatrix} = \begin{pmatrix} 0 & 1 \\ 1 & 0 \end{pmatrix}$$

The 32 products presented above as generating the negation operator $\urcorner$ employ 32 different operators, among which there are 28 operators of negative logic and 4 operators of positive logic.

Consecutive application of appropriate pairs of these operators achieves complete negation, whereas each operator alone can be considered as a subnegator performing certain forms of partial negation. In this ways we introduce "degrees" of negation and enlarge the negating power of the complete logic set 32 times, in comparison with the single negator of positive logic. The class of negators of negative logic is characterized by an interesting ambivalence: each of them can participate in full negation, if joined by a corresponding subnegator, or in affirmation, if joined by a corresponding inverse.

Generalized complementation

We have arrived at the need to consider negative logic operations as a result of a generalization of the notion of a logic inverse which became possible due to the matrix formulation of logic. This is the most logical and consistent approach, which is not initiated by hazard but is forced upon us by the nature of the inverse operation in matrix algebra.

However, there are other ways of developing the negative extension of logic. One of them is based on a generalization of the complement operation, which in positive logic was defined with respect to the tautology operator $\hat{1}$:

$$\overline{L_i} = \hat{1} - L_i \,, \quad L_i \in \Omega^+ \qquad i = 1, 16$$

If instead of the complementation with respect to the tautology operator $\hat{1}$ we consider a general form of complementation

$$L_i - L_j = L_k \,, \quad L_i, L_j \in \Omega^+ \qquad i,j = 1, 16$$
$$L_k \in \Omega^- \qquad k = 1, 66$$

we can then generate the complete set of 66 operators of negative logic, by considering all possible generalized complementations and leaving only those negative operators which are different.

Whereas Ω^+ -logic was generated from Ω^- -logic by means of multiplication, conversely Ω^- -logic can be generated from Ω^+ -logic by subtraction. Any negative operator L_k^- can be expressed via a generalized complementation of two positive operators

$$L_k^- = L_i^+ - L_j^+$$

with the corresponding bilinear product taking the form:

$$\langle x \mid L_k^- \mid y\rangle = \langle x \mid L_i^+ - L_j^+ \mid y\rangle = \langle x \mid L_i^+ \mid y\rangle - \langle x \mid L_j^+ \mid y\rangle \,.$$

Since

$$L_i^+ - L_j^+ = \overline{L_j^+} - \overline{L_i^+}$$

the bilinear product can also be expressed as

$$<x \mid L_k^- \mid y> = \overline{<x \mid L_j \mid y>} - \overline{<x \mid L_i \mid y>} .$$

In terms of generalized complementations negative logic decays into a pair of two positive logics bonded between themselves by subtraction.

Though the problem of negative logic has decayed into two problems of positive logic, due to subtraction, the final outcome of evaluation is, nevertheless, still defined in the ternary set of matrix logic:

$$E_3 = \{-1, 0, 1\}.$$

Multiplication Lie

Finally, there is a third way to introduce and develop negative matrix logic. In contrast to the previous two, it is based not on logic but on pure mathematical considerations involving a special operation of matrix algebra: multiplication Lie.

Multiplication Lie, or the commutator of operators, is a operation defined as follows

$$[L_i,L_j] = -[L_j,L_i] = L_iL_j - L_jL_i ,$$

where rectangular brackets are commonly used to denote the operation.

We say that two operators L_i and L_j commute if

$$[L_i,L_j] = 0,$$

and do not commute if

$$[L_i,L_j] \neq 0 .$$

Application of multiplication Lie to operators of the set Ω^+ shows that positive logic consists mostly of non-commuting operators. Though this fact is important on its own merits, our concern is here not with the commutation of logic operators but with the fact that products of multiplication Lie, in case of non-commutation, can only be negative logic operators.

Consider several important examples:

$$[\wedge,\vee] = \perp - \dashv = \begin{pmatrix} 0 & -1 \\ 1 & 0 \end{pmatrix}$$

$$[\uparrow,\downarrow] = \top - \vdash = \begin{pmatrix} 0 & 1 \\ -1 & 0 \end{pmatrix}$$

$$[\rceil,\rightarrow] = \vee - \downarrow = \begin{pmatrix} -1 & 0 \\ 0 & 1 \end{pmatrix}$$

$$[\rightarrow,\backslash] = \vdash - \perp = \begin{pmatrix} 1 & 0 \\ 0 & -1 \end{pmatrix}$$

On the set Ω^+ multiplication Lie manifests itself in generating either the empty operator $\hat{0}$ or negative logic operators. Continuing this line of reasoning and joining multiplication Lie with other matrix operations, we can generate the complete set of negative logic Ω^-.

Negative logic vectors

Finally, negative logic can be introduced in purely axiomatic fashion. Because matrix logic is constructed over vector space, the 3rd axiom of vector space is a sufficient mathematical argument obliging us to consider the negative logic vectors:

$$(0,-1) \quad \text{and} \quad (-1,0)\,,$$

with a reciprocal relation between them, similar to the reciprocal relation of logic vectors in positive logic.

In any event, even without being defined from the beginning, negative logic vectors ought to appear as the result of the application of L^--operators which could carry a negative sign from a matrix element or elements into the resulting product vector. The extension of logic attributable to the introduction of the negative values can be described in terms of the following truth-lattices:

Positive logic

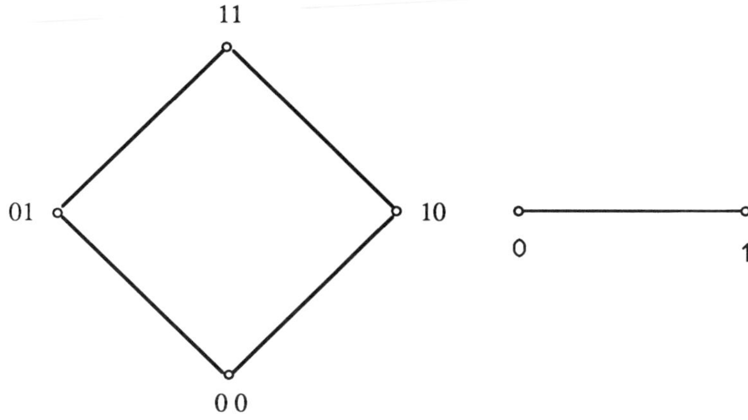

Logic vectors Logic scalars

Negative logic

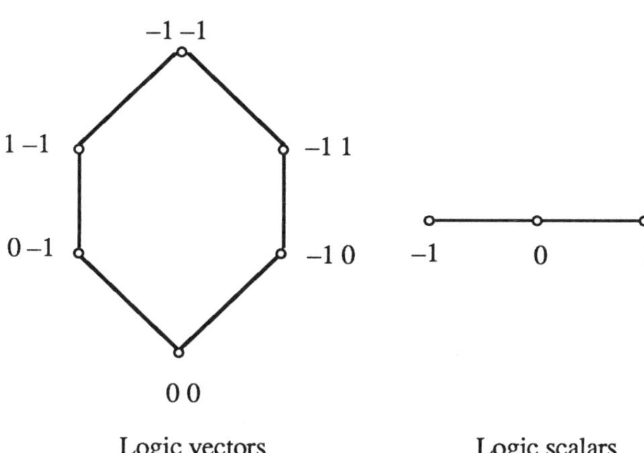

Logic vectors Logic scalars

Whereas the number of scalar truth-values in negative logic increases, by comparison with positive logic, from 2 to 3, the number of logic vectors grows from 4 to 6, and the number of logic functions from 16 in positive logic to 81 in complete logic $\Omega^+ \cap \Omega^-$.

Our prime concern here is with the negative logic vectors. As in positive logic, where we defined the vectors (0,1) and (1,0) as the base vectors and (0,0), (1,1) as derivative ones, in negative logic we also consider –(0,1) = (0,–1) and –(1,0) = (–1,0) as the base vectors. The remaining four vectors

(0,0), (–1,1) and (1,–1), (–1,1)

are the derivative vectors, which appear during intermediary steps of computation.
The operators of negative logic and the operators of positive logic act identically in the spaces of negative and positive base logic vectors:

$$<x \mid \overset{+}{L} \mid y> = <-x \mid \overset{+}{L} \mid -y> ,$$
$$<x \mid \overset{-}{L} \mid y> = <-x \mid \overset{-}{L} \mid -y> .$$

with the results of the first row defined on the binary set $E_2 = \{0,1\}$, and the results of the second row defined on the ternary set $E_3 = \{-1, 0, 1\}$. In this case negative logic remains negative and positive logic remains positive, with the outcome being dependent not on the operators used but on the space in which they act.
However, if we consider the cross-interaction of positive and negative logic vectors where the logic operator acts between negative and positive logics the situation becomes more complex. A positive logic operator L^+acting between these two logics will produce one of the negative logic functions, defined on the set $E_2^{(-)}$:

$$<x \mid \overset{+}{L} \mid -y> = <x \mid \overset{-}{L} \mid y> ,$$

with

$$\overset{-}{L} = -\overset{+}{L} .$$

But application of a negative logic operator L^- in the same circumstances can result either in a positive logic function or in a negative logic function, depending which operator from the set Ω^- is used:

$$<x \mid \overset{-}{L_i} \mid - y> = \begin{cases} <x \mid \overset{+}{L_i} \mid y> , & i = 1, 16. \\ <x \mid \overset{-}{L_j} \mid y> , & j = 1, 50. \end{cases}$$

The asymmetry of the relationships between positive and negative logic is apparent in fundamental products, where a product containing or even consisting entirely of negative logic functions actually represents a positive logic function. Consider for example:

$$<x \mid \vee \mid y><x \mid \overset{-1}{\vee} \mid y> = \overline{<x \mid y>}$$
$$<x \mid \overset{-1}{\rightarrow} \mid y><x \mid \overset{-1}{\leftarrow} \mid y> = <x \mid y>$$

Negative logic, being richer in possibilities, allows a more subtle analysis of logic functions in general and especially of such an asymmetric logic function as implication.
Consider the behaviour of implication and its inverse, which is implication in negative logic:

$\Omega^{+}: \rightarrow$	$\Omega^{-}: \overset{-1}{\rightarrow}$
$(1\ \ 0)\begin{pmatrix}1 & 1\\ 0 & 1\end{pmatrix}\begin{pmatrix}1\\ 0\end{pmatrix} = 1$	$(-1\ \ 0)\begin{pmatrix}1 & -1\\ 0 & 1\end{pmatrix}\begin{pmatrix}-1\\ 0\end{pmatrix} = 1$
$(1\ \ 0)\begin{pmatrix}1 & 1\\ 0 & 1\end{pmatrix}\begin{pmatrix}0\\ 1\end{pmatrix} = 1$	$(-1\ \ 0)\begin{pmatrix}1 & -1\\ 0 & 1\end{pmatrix}\begin{pmatrix}0\\ -1\end{pmatrix} = -1$
$(0\ \ 1)\begin{pmatrix}1 & 1\\ 0 & 1\end{pmatrix}\begin{pmatrix}1\\ 0\end{pmatrix} = 0$	$(0\ \ -1)\begin{pmatrix}1 & -1\\ 0 & 1\end{pmatrix}\begin{pmatrix}-1\\ 0\end{pmatrix} = 0$
$(0\ \ 1)\begin{pmatrix}1 & 1\\ 0 & 1\end{pmatrix}\begin{pmatrix}0\\ 1\end{pmatrix} = 1$	$(0\ \ -1)\begin{pmatrix}1 & -1\\ 0 & 1\end{pmatrix}\begin{pmatrix}0\\ -1\end{pmatrix} = 1$

The signatures of both implications are the same except for one row which in Ω^{+} -logic is 1 but in Ω^{-} -logic becomes –1. Observe that the row in question corresponds to the notorious case in which true is allowed to follow from false. In this case,which frequently causes confusion in binary logic, the truth-value 0 cannot be assigned since then it would not be possible to distinguish between implication as conditional and equivalence as biconditional. But assigning to it the truth-value –1 does makes the difference without either transforming conditional into biconditional, or denying the validity of implication, at the same time weakening the implication of pure positive logic.

3.3 Continuous logic Ω^{ε}

The introduction of matrix operators in place of logic connectives and the interpretation of the logical process as the transformation of matrix operators are important not only on their own merits. Negative logic, developed in the preceding section, clearly demonstrates that matrix operator algebra serves to provide unambiguous mathematical means for extensions of logic which otherwise can be conceived only in a postulatory manner, if at all.In addition to positive and negative logics, considered in this chapter, we shall now present another important extension of logic into the domain of fractional and continuous logic values. Conceptually, continuous logic can be regarded as an extension which deals with approximate or imprecise forms of reasoning, that is, reasoning which involves partially true and partially false truth-values.

Since the numerical measures of true and false are given by the integers 1 and 0 respectively, it seems quite natural to associate partially true and partially false with non-integer or fractional truth-values. In the context of a matrix formulation of logic this innovation not only makes good theoretical sense as a clarification of approximate reasoning but it has important practical advantages, too. Whenever the elements of a matrix are in fractional form, we can carry out computations using fractional forms. However, when such computation is processed by the computer it is normally carried in decimal form. For example, the computer word for 1/3 might be 0.3333333333. As a result, the product of a matrix and the computer form of its inverse will not always result in precisely the identity matrix. Within the computer the product might appear for example as

$$\begin{pmatrix} 0.9998 & 0.0001 \\ 0.0002 & 1.0001 \end{pmatrix},$$

which is not the identity matrix, but is sufficiently close to be considered as such.

In addition to considerations of a general nature, such examples have made apparent the importance of developing the extension of matrix logic into the domain of fractional truth-values.

Strong normalization

There are several equivalent ways of developing this type of extension. In one of them we begin with fractional logic vectors but retain integer elements for matrix logic operators. We consider logic vectors with components $x_i \geq 0$, satisfying the strong normalization condition:

$$\sum x_i = 1, \quad 0 \leq x_i \leq 1.$$

In the other approach, we retain the logic vectors in their usual form (0,1) and (1,0), so that $x_i \in E_2$, but introduce fractional values for the elements of the logic operators:

$$0 \le a_{ij} \le 1 \ ,$$

with

$$0 \le \langle x \mid A \mid y \rangle \le 1 \ ,$$

where A denotes a non-integer logic operator.

Each operator A has its nearest corresponding logic operator L:

$$A_i - L_i = \begin{pmatrix} \pm\varepsilon & \pm\varepsilon \\ \pm\varepsilon & \pm\varepsilon \end{pmatrix}$$

where $\pm\,\varepsilon << 0$ varies and can be made as small as necessary.

This implies that we have replaced the matrix-operator components 0 and 1 with ε and $1-\varepsilon$ respectively. Thus, in the neighbourhood of any logic operator L_i, there is an infinite number of fractional operators A_i.

Fractional matrix logic, as defined above, begins to display fundamental properties of adaptivity and that is why we shall define it as an adaptive logic. This becomes evident if we consider the iterative application of the A-operators. In the set Ω^+ we have the idempotent subset

$$\{\leftrightarrow, \wedge, \uparrow, \perp, \top, \dashv, \vdash\};$$

whenever, for any of these operators, $L^n = L$, their iterative application to a logic vector does not modify the result of the first application.

In adaptive logic none of the iterations reproduces the initial A-operator: each iteration adaptively transforms the A-operator into another A-operator. There are three distinct outcomes of iterative operations on the set of adaptive A-operators: one subset indefinitely modifies itself; another subset converges to the empty operator $\hat{0}$, if the number of iterations n is sufficiently large, $n \to \infty$, and ε is sufficiently small, $\varepsilon << 0$; and the third and most important subset of A-operators which tend to some stationary limits. Some properties of adaptive logic operations very closely resemble certain properties of multivalued and intuitionistic logic where we may recall cases when

not not $x \neq x$ – multivalued logic

not not x is weaker than x – intuitionistic logic.

Adaptive equivalence and negation

We have seen the fundamental role played in matrix logic by the only orthogonal operators of equivalence and negation: $\{\leftrightarrow, \urcorner\}$. It is quite remarkable that the set of adaptive equivalence and negation in adaptive logic

$$\leftrightarrow_A = \begin{pmatrix} 1-\varepsilon & \varepsilon \\ \varepsilon & 1-\varepsilon \end{pmatrix}; \quad \urcorner_A = \begin{pmatrix} \varepsilon & 1-\varepsilon \\ 1-\varepsilon & \varepsilon \end{pmatrix}$$

possesses exceptional properties, making it a fundamental basis of adaptive logic.
One fundamental property of A-equivalence and A-negation are normalization conditions

$$\sum_{i=1}^{2} a_{ij} = 1 \quad ; \quad \sum_{j=1}^{2} a_{ij} = 1 \ ,$$

which do not hold for any other A-operators, and which together with the normalization condition for the adaptive logic vectors, with fractional components $1-\varepsilon$ and ε:

$$1-\varepsilon+\varepsilon = 1$$

complete the picture, and make the adaptive logic of $\leftrightarrow_A$ and $\urcorner_A$ fully consistent.
Another fundamental property of $\leftrightarrow_A$ and $\urcorner_A$ is that they form a commuting set:

$$[\leftrightarrow_A, \urcorner_A] = 0,$$

implying that

$$\leftrightarrow_A \urcorner_A = \urcorner_A \leftrightarrow_A.$$

Furthermore, at the same level of uncertainty ε, the products $\leftrightarrow_A \urcorner_A$ and $\urcorner_A \leftrightarrow_A$ remain normalized:

$$\begin{pmatrix} 1-\varepsilon & \varepsilon \\ \varepsilon & 1-\varepsilon \end{pmatrix} \begin{pmatrix} \varepsilon & 1-\varepsilon \\ 1-\varepsilon & \varepsilon \end{pmatrix} = \begin{pmatrix} 2\varepsilon(1-\varepsilon) & (1-\varepsilon)^2+\varepsilon^2 \\ \varepsilon^2+(1-\varepsilon)^2 & 2\varepsilon(1-\varepsilon) \end{pmatrix}$$

$$\begin{pmatrix} \varepsilon & 1-\varepsilon \\ 1-\varepsilon & \varepsilon \end{pmatrix} \begin{pmatrix} 1-\varepsilon & \varepsilon \\ \varepsilon & 1-\varepsilon \end{pmatrix} = \begin{pmatrix} 2\varepsilon(1-\varepsilon) & \varepsilon^2+(1-\varepsilon)^2 \\ (1-\varepsilon)^2+\varepsilon^2 & 2\varepsilon(1-\varepsilon) \end{pmatrix}$$

with $2\varepsilon(1-\varepsilon) + (1-\varepsilon)^2 + \varepsilon^2 = (1-\varepsilon+\varepsilon)^2 = 1$.

Probably, the most puzzling result of adaptive logic is that at infinity A-equivalence and A-negation are not distinguishable. While in positive integer logic

$$\leftrightarrow^n = \leftrightarrow$$

$$\urcorner^n = \begin{cases} \leftrightarrow & \text{for n even} \\ \urcorner & \text{for n odd} \end{cases}$$

in normalized adaptive logic:

$$\leftrightarrow^n_A \underset{n \to \infty}{\rightarrow} \frac{1}{2}(\hat{1}) \ ,$$

$$\urcorner^n_A \underset{n \to \infty}{\rightarrow} \frac{1}{2}(\hat{1}) \ .$$

At infinity adaptive logic converges to equilibrium with maximal entropy, whatever were the initial states of the normalized logic vectors:

$$\langle x| \leftrightarrow^n_A |y\rangle \underset{n \to \infty}{\rightarrow} \frac{1}{2}; \qquad \langle x| \urcorner^n_A |y\rangle \underset{n \to \infty}{\rightarrow} \frac{1}{2} .$$

This not only indicates that the fundamental set of truth-values of adaptive logic is

$$\{0, 1/2, 1\}$$

but also, since true and false become equiprobable, that A-equivalence and A-negation become indistinguishable at infinity.
Under finite conditions, $\leftrightarrow_A$ and $\urcorner_A$, even with different degrees of uncertainty, both act as identity operators on the equiprobable logic vectors:

$$\left(\frac{1}{2} \ \frac{1}{2}\right) \leftrightarrow_A = \left(\frac{1}{2} \ \frac{1}{2}\right)$$

$$\left(\frac{1}{2} \ \frac{1}{2}\right) \urcorner_A = \left(\frac{1}{2} \ \frac{1}{2}\right) .$$

Comparison of logic vectors in different extensions of matrix logic gives some insights into their differences and similarities. Denoting the components of logic vector |x> with p and q we observe the following relationships:

$$\Omega^+: \quad p+q=1 \qquad \langle x \mid x \rangle = 1$$
$$\Omega^-: \quad p+q=-1 \qquad \langle x \mid x \rangle = 1$$
$$\Omega^\varepsilon: \quad p+q=1 \qquad \langle x \mid x \rangle = 1-2\varepsilon+2\varepsilon^2$$

In normalized logics special interest attaches to the singular negators $\{\perp, \top, \dashv, \vdash\}$. These offer an exit from fractional adaptive logic into matrix logic with definite truth-values (0 1) and (1 0).

For left logic:

$$\begin{pmatrix} 0 & 0 \\ 1 & 1 \end{pmatrix} \begin{pmatrix} p \\ q \end{pmatrix} = \begin{pmatrix} 0 \\ p+q \end{pmatrix} = \begin{pmatrix} 0 \\ 1 \end{pmatrix}$$

$$\begin{pmatrix} 1 & 1 \\ 0 & 0 \end{pmatrix} \begin{pmatrix} p \\ q \end{pmatrix} = \begin{pmatrix} p+q \\ 0 \end{pmatrix} = \begin{pmatrix} 1 \\ 0 \end{pmatrix}$$

For right logic:

$$(p \;\; q) \begin{pmatrix} 0 & 1 \\ 0 & 1 \end{pmatrix} = (0 \;\; p+q) = (0 \;\; 1)$$

$$(p \;\; q) \begin{pmatrix} 1 & 0 \\ 1 & 0 \end{pmatrix} = (p+q \;\; 0) = (1 \;\; 0) \,.$$

Weak normalization

So far we have considered adaptive logic with normalized logic vectors

$$p+q=1,$$

and with two normalized operators $\{\leftrightarrow_A, \urcorner_A\}$.

We now shall weaken the normalization condition and allow that

$$p+q \leq 1.$$

This weak adaptive logic gives us a new and very interesting possibility. The set of logic operators satisfying the condition of weak normalization is

$$\{\Omega^{\varepsilon}_{weak}\} = \{\, \leftrightarrow, \rceil, \wedge, \setminus, /, \uparrow, \hat{0} \,\}.$$

Apart from these seven operators all other operators can violate weak normalization, producing vectors with

$$p + q > 1$$

which have no logic meaning and therefore must be excluded.
The multiplication table of weak adaptive logic is thus reduced to:

	↑	\	/	∧	⌉	↔	0
↑	↑	0	/	0	/	↑	0
\	\	0	∧	0	∧	\	0
/	0	↑	0	/	↑	/	0
∧	0	\	0	∧	\	∧	0
⌉	\	↑	∧	/	↔	⌉	0
↔	↑	\	/	∧	⌉	↔	0
0	0	0	0	0	0	0	0

with the valuation device

$$0 \le \langle x \,|\, L_i \,|\, y \rangle \le 1 ; \quad i = 1,7 ,$$

being defined across the unit interval, including its lower and upper bounds.

Continuum of truth–values

In continuous logic the binary alphabet of truth-values 1 and 0 is replaced by a continuum of truth-values $0 \le p, q \le 1$.In the modal interpretation p and q can be viewed as probability:

p – as the probability of finding a logic vector in the true state or state (0,1);

and

q – as the probability of finding a logic vector in the false state or state (1,0).

The probability that the logic vector is either in the true or in the false state is 1, so that the sum of probabilities $p + q = 1$, the strong normalization condition.
Hence, in continuous logic, logic vectors can be represented as the vector sum of two probabilities p and q, while the set of all possible states corresponds to a unit circle in the phase space of truth-values:

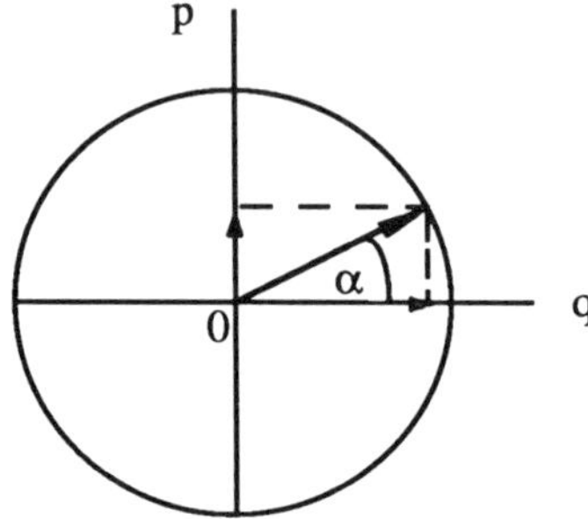

This approach reveals yet another aspect of logic, making it possible to interpret literally the bra and ket logic spaces as adjoint conjugate transposes in the Dirac sense:

$$\langle \mathrm{bra} | = | \mathrm{k\tilde{e}t} \rangle^{*}$$

and

$$\langle x \mid y \rangle = \langle y \mid x \rangle^{*}.$$

Logic vectors in continuous logic acquire the genuine properties of wave processes.
In the complex unit circle each bra logic vector is multiplied by the phase factor $e^{-i\alpha}$ and each ket logic vector by the phase factor $e^{i\alpha}$, where α is a periodic parameter:

$$\alpha \pm 2\pi n = \alpha , \qquad n = 0,1,2...$$

Consequently, in the phase space of continuous logic each point can be associated with infinitely many logic vectors of the same truth status:

$$(1,0)\ e^{-i\alpha} = (1,0)\ e^{-i(\alpha + 2\pi n)}$$

$$(0,1)\ e^{-i\alpha} = (0,1)\ e^{-i(\alpha \pm 2\pi n)}$$

Though intermediary logic operations involve complex values, these do not appear in the final results of valuation, which are expressed in real numbers. Consider for example the disjunction function in continuous logic:

$$\begin{pmatrix} e^{-i\alpha} & 0 \end{pmatrix} \begin{pmatrix} 0 & 1 \\ 1 & 1 \end{pmatrix} \begin{pmatrix} e^{i\alpha} \\ 0 \end{pmatrix} = \begin{pmatrix} 0 & e^{-i\alpha} \end{pmatrix} \begin{pmatrix} e^{i\alpha} \\ 0 \end{pmatrix} = 0$$

$$\begin{pmatrix} e^{-i\alpha} & 0 \end{pmatrix} \begin{pmatrix} 0 & 1 \\ 1 & 1 \end{pmatrix} \begin{pmatrix} 0 \\ e^{i\alpha} \end{pmatrix} = \begin{pmatrix} 0 & e^{-i\alpha} \end{pmatrix} \begin{pmatrix} 0 \\ e^{i\alpha} \end{pmatrix} = 1$$

$$\begin{pmatrix} 0 & e^{-i\alpha} \end{pmatrix} \begin{pmatrix} 0 & 1 \\ 1 & 1 \end{pmatrix} \begin{pmatrix} e^{i\alpha} \\ 0 \end{pmatrix} = \begin{pmatrix} e^{-i\alpha} & e^{-i\alpha} \end{pmatrix} \begin{pmatrix} e^{i\alpha} \\ 0 \end{pmatrix} = 1$$

$$\begin{pmatrix} 0 & e^{-i\alpha} \end{pmatrix} \begin{pmatrix} 0 & 1 \\ 1 & 1 \end{pmatrix} \begin{pmatrix} 0 \\ e^{i\alpha} \end{pmatrix} = \begin{pmatrix} e^{-i\alpha} & e^{-i\alpha} \end{pmatrix} \begin{pmatrix} 0 \\ e^{i\alpha} \end{pmatrix} = 1$$

The wave properties of logic operations introduce an important new factor into logic analysis and a large class of theoretical and applied problems can be formulated in terms of the framework of continuous logic.

3.4 Hypercube logic Ω^n

We shall conclude this chapter with the n-dimensional generalization of matrix logic, or hypercube logic.

We define hypercube logic over the bra and ket spaces of hyper-Boolean logic vectors:

$$(x_1, x_2, x_3, \ldots x_n), \quad x_i \in E_2$$

with

$$\sum_{i=1}^{n} x_i = 1, \quad \langle x \mid x \rangle = 1, \quad \langle x \mid y \rangle = \delta_{xy}\,.$$

Among n components of a hyper-Boolean vector only one is equal to 1, with all remaining n–1 components equal to 0. A complete set of n hyper-Boolean logic vectors forms the orthonormal basis which spans the n-dimensional vector space:

$$
\begin{matrix}
(1 & 0 & 0 & \ldots & 0) \\
(0 & 1 & 0 & \ldots & 0) \\
(0 & 0 & 1 & \ldots & 0) \\
. & . & . & \ldots & . \\
(0 & 0 & 0 & \ldots & 1)
\end{matrix}
$$

The hyper-Boolean vectors can also be associated with the vertices of an n-dimensional Boolean cube, which is why we define the n-dimensional generalization as hypercube logic. The 2-dimensional bra and ket logic vectors which we have considered so far thus represent the simplest possible set of hyper-Boolean logic vectors. Just as hyper-Boolean logic vectors are an n-dimensional generalization of 2-dimensional ones, the hypercube logic operators are an $(n \times n)$ generalization of (2×2) matrix logic operators. Below we present an (8×8) example of a hypercube matrix operator, which is characterized by homunculus type architecture, and which will later be denoted as an H-operator:

$$
H = \begin{pmatrix}
\begin{pmatrix} (L_{11}) & (L_{12}) \\ (L_{21}) & (L_{22}) \end{pmatrix} & \begin{pmatrix} (L_{13}) & (L_{14}) \\ (L_{23}) & (L_{24}) \end{pmatrix} \\
\begin{pmatrix} (L_{31}) & (L_{32}) \\ (L_{41}) & (L_{42}) \end{pmatrix} & \begin{pmatrix} (L_{33}) & (L_{34}) \\ (L_{43}) & (L_{44}) \end{pmatrix}
\end{pmatrix}
$$

The H-operator is a square matrix which can be partitioned into a set of $(n/2 \times n/2)$ suboperators, which could in their own turn be partitioned into a set of $(n/4 \times n/4)$ suboperators, and so on until the H-operator is partitioned into a set of basic (2×2) logic operators L_{ij}. It is worthy of special remark that if n runs to infinity, the above reduction takes the form of an infinite regress, and hypercube logic becomes Hilbert space logic.

On the set of hypercube operators $H_i \in \Omega^n$ we require that for any pair of hyper-Boolean vectors $\langle x|$ and $|y\rangle$:

$$
\langle x \mid H_i \mid y\rangle \in E_2 \quad , \quad E_2 = \{0,1\} \ ,
$$

and for any pair of hypercube operators $H_i, H_j \in \Omega^n$ their product $H_iH_j = H_k$ is again a hypercube operator:

$$H_k \in \Omega^n.$$

Particular interest attaches to normalized H-operators:

$$\sum_{i=1}^{n} h_{ij} = 1 \quad \text{and} \quad \sum_{j=1}^{n} h_{ij} = 1,$$

where h_{ij} are matrix elements of H.

This condition, together with need to prevent the appearance of forbidden interactions, reduces the set of possible L-operators which can be employed in hypercube logic to the fundamental base of adaptive logic:

$$\{\leftrightarrow, \rceil, \wedge, /, \backslash, \uparrow, \hat{0}\}.$$

In an n-dimensional logic vector space the number of different hypercube operators is given by n!.

CHAPTER 4

MATRIX LOGIC AND FUNDAMENTAL PHYSICS

A variety of mathematical methods are employed in solving diverse problems of quantum theory and fundamental physics. With the development of relativity and quantum theories, an observer has become an indispensable component of the measurement procedure, shifting the interest of physicists towards the investigation of the information properties of matter.
The Heisenberg uncertainty principle, the Bohr complementarity principle and the Cooper thesis of quantum mechanics clearly have the character of information principles. The Pauli exclusion postulate is precisely what in the domain of logic takes the form of the law of Excluded Middle:

$$\langle x \mid \vee \mid \bar{x} \rangle = 1 .$$

On the other hand, the Gödel Incompletness Theorem can in the domain of quantum physics be perceived as a kind of a quantum-mechanical statement.
The matrix operator formulation of logic opens new avenues for the application of logic to the formalization and study of physical phenomena. Just as Boolean logic algebra bridges logic and classical physics, matrix logic algebra can serve as a unifying link between fundamental physics and logic. One of the key reasons for such an assumption is that both quantum physics and matrix logic are built around a similar mathematical construct. On one hand we have seen that the bilinear product:

$$\langle x \mid L \mid y \rangle ,$$

where $\langle x|$ and $|y\rangle$ are logic vectors from two adjoint vector spaces, and L a logic operator, is a basic valuation device of logic. On the other hand, according to the formalism developed by Dirac, precisely the same measuring device lies at the foundation of modern quantum physics:

$$\langle \varphi^* \mid Q \mid \psi \rangle ,$$

where φ^* and ψ are the state vectors or wave functions from two adjoint dual spaces, and Q a quantum-mechanical operator. The fact that such a perfect correspondence exists between the basic measurement devices of quantum physics and logic has far-reaching implications for

both fields.
In quantum physics the wave functions do not represent an observable quantity; only the square of a modulo of a wave function has physical meaning:

$$\|\psi\|^2 = \langle\psi^* \mid \psi\rangle$$

where ψ^* is the conjugate transpose of ψ.

Similarly, in matrix logic, the logic vectors also represent "logically" unobservable quantities, yielding observable values through the application of the bilinear valuation device. To measure a physically observable quantity or to infer a logically observable quantity, we have to introduce a corresponding operator between the two adjoint bra and ket vectors. As a matter of fact these can be viewed within two alternative frames of reference: in the quantum domain as state vectors, and in the domain of logic as logic vectors.
To elaborate, what is described in quantum language as the collapse of the state vector represents, in the domain of logic, a decision process.

In the Heisenberg formulation of quantum mechanics, the transition spectral operators are defined in such a way that their products do not introduce new frequencies beyond those which were already in the multiplying operators. In a similar way, multiplications of logic operators are constrained by the set of forbidden interactions. As a result no other operators besides those given in the initial set are allowed to appear.
In quantum physics an act of measurement, as described by the bracket closure $\langle\Psi^* \mid Q \mid \Psi\rangle$, projects a pure state out of the mixed state of a quantum system, whereas the bilinear logic closure also extracts a sharp scalar truth-value from a logic system which is capable of being in intermediary undecided states.
Thus we are given two alternative interpretations: the measurement act in the quantum domain can equally be viewed as a decision process in the domain of logic. In accordance with the general scheme for translating between the two interpretations, what is seen from the logical view as an inference process can be described in quantum language as the collapse of the state vector.

In view of the links described above and other similarities which will be presented later, one cannot escape the impression that in logic and in quantum physics we are dealing with different modes of one fundamental phenomenon. Could it be that matrix logic is actually revealing the logical processes to be quantum-mechanical in nature? Is the similarity discovered here superficial or does it reflect a fundamental relation? And if the latter is true, can we derive or establish an explicit connection between logic and quantum-physical processes? If the codes of logic and the codes of matter have the same structure, which quantum-mechanical processes and properties can be held responsible for the form that logic takes?

4.1 Spinor logic

A complete definition of a state of a particle requires not only a specification of its space-time coordinates but also of the direction of the spin vector. Consequently, the wave function of a particle becomes dependent on the spin variable which can be assigned to it as an index, specifying the direction of spin, either up or down.
With the exception of spin, all quantum-mechanical quantities have corresponding macro analogues in classical physics. The investigation of spin in the context of matrix logic puts this unresolved problem into a new perspective. The arbitrary spin state is represented by a spinor or two-component wave function, quantized into spin-up $\psi_\uparrow$ and spin-down $\psi_\downarrow$ components:

$$\psi = \begin{pmatrix} \psi_\uparrow \\ \psi_\downarrow \end{pmatrix}.$$

A spinor is arbitrary to the extent of a phase multiplier. Squares of modulo $|\psi_\uparrow|^2$ and $|\psi_\downarrow|^2$ give the probabilities of observing transitions into the states $\psi_\uparrow$ and $\psi_\downarrow$ respectively, and the sum of the probabilities is unity:

$$|\psi_\uparrow|^2 + |\psi_\downarrow|^2 = 1.$$

Spin operators may have two eigenvalues and are expressed by the set of Pauli matrices, which together with the identity matrix form the unitary group of symmetry SU(2):

$$\sigma_0 = \begin{pmatrix} 1 & 0 \\ 0 & 1 \end{pmatrix} \quad \sigma_x = \begin{pmatrix} 0 & 1 \\ 1 & 0 \end{pmatrix} \quad \sigma_y = i\begin{pmatrix} 0 & -1 \\ 1 & 0 \end{pmatrix} \quad \sigma_z = \begin{pmatrix} 1 & 0 \\ 0 & -1 \end{pmatrix},$$

with cyclic closure

$$\sigma_x\sigma_y = -i\sigma_z, \quad \sigma_y\sigma_z = -i\sigma_x, \quad \sigma_z\sigma_x = -i\sigma_y,$$

anticommutation

$$\sigma_x\sigma_y + \sigma_y\sigma_x = 0$$

$$\sigma_y\sigma_z + \sigma_z\sigma_y = 0$$

$$\sigma_z\sigma_x + \sigma_x\sigma_z = 0 ,$$

and

$$\sigma_x^2 = \sigma_y^2 = \sigma_z^2 = \sigma_0 .$$

Spinor transformations occur in unitary spinor space with the observed quantities being expressed by the matrix elements $<\varphi \mid f(\sigma) \mid \psi>$ where φ and ψ are spinors, and $f(\sigma)$ is a function of spin operators.

Any transformation which leaves $<\varphi \mid f(\sigma) \mid \psi>$ unchanged has no physical manifestation, and any transformation of the form

$$\Psi' = Q\Psi, \ \sigma' = Q\sigma Q^{-1}$$

where Q is a unitary matrix, is possible.

A link can be established between the spin formalism and matrix logic. In the σ_z-representation two eigenstates of spinor have the most simple forms

$$\psi_\uparrow = \begin{pmatrix} 1 \\ 0 \end{pmatrix}; \qquad \psi_\downarrow = \begin{pmatrix} 0 \\ 1 \end{pmatrix},$$

which are precisely the binary vectors we situated at the foundation of the matrix logic If we introduce the rule

$$|x> = -|\bar{x}> ,$$

similar to the rule of the conjugation of a complex unit: $i = -i^*$, then the Pauli spinor matrices

$$\begin{pmatrix} 1 & 0 \\ 0 & 1 \end{pmatrix} \quad \begin{pmatrix} 0 & 1 \\ 1 & 0 \end{pmatrix} \quad \begin{pmatrix} 0 & -1 \\ 1 & 0 \end{pmatrix} \quad \begin{pmatrix} 1 & 0 \\ 0 & -1 \end{pmatrix}$$

will act upon spinors in σ_z-representation in the same way as operators of a strictly binary logic base

$$\begin{pmatrix} 1 & 0 \\ 0 & 1 \end{pmatrix} \quad \begin{pmatrix} 0 & 1 \\ 1 & 0 \end{pmatrix} \quad \begin{pmatrix} 0 & 0 \\ 1 & 1 \end{pmatrix} \quad \begin{pmatrix} 1 & 1 \\ 0 & 0 \end{pmatrix}$$

act upon logic vectors in the ket space. Naturally, for the logic vectors of bra space, the matrices of the binary logic base must be transposed:

$$\begin{pmatrix} 1 & 0 \\ 0 & 1 \end{pmatrix} \quad \begin{pmatrix} 0 & 1 \\ 1 & 0 \end{pmatrix} \quad \begin{pmatrix} 0 & 1 \\ 0 & 1 \end{pmatrix} \quad \begin{pmatrix} 1 & 0 \\ 1 & 0 \end{pmatrix} .$$

The fundamental novelty of spinor behaviour lies in the absence of identity with respect to 2π revolution, which produces a factor -1:

$$\psi(2\pi) = \begin{pmatrix} \cos\frac{\theta}{2} & i\sin\frac{\theta}{2} \\ i\sin\frac{\theta}{2} & \cos\frac{\theta}{2} \end{pmatrix} \psi(0) = -\,\psi(0)$$

Consequently, a spinor can be characterized by a ternary balanced set

$$\{-1, 0, 1\} ,$$

which is precisely the balanced set of scalar truth-values of the complete logic set.

In the view that spin is the only quantum-mechanical quantity which is not derivable from the classical notions and for which the macroanalogue remains unknown it is not excluded that binary logic, extended into the domain of negative values, can serve as such a macroanalogue. In such an event, which is indeed probable, it might be possible not only to draw parallels between the logic processes and the spinor transformations, but to relate both in a fundamental sense.

4.2 Relativistic spinor logic

The spin formalism uses the set of Pauli matrices

$$\{\sigma_0, \sigma_x, \sigma_y, \sigma_z\}$$

where σ_x, σ_y, σ_z are associated with the corresponding Cartesian space axis xyz. From the point of view of logic, we may say that it involves the subset of positive logic:

$$\sigma_0 = \begin{pmatrix} 1 & 0 \\ 0 & 1 \end{pmatrix} \qquad \sigma_x = \begin{pmatrix} 0 & 1 \\ 1 & 0 \end{pmatrix}$$

and the subset of negative logic:

$$\sigma_y = i \begin{pmatrix} 0 & -1 \\ 1 & 0 \end{pmatrix} \qquad \sigma_z = \begin{pmatrix} 1 & 0 \\ 0 & -1 \end{pmatrix}$$

with commutation holding for the positive subset,

$$\sigma_0 \sigma_x - \sigma_x \sigma_0 = 0,$$

and anticommutation holding for the negative subset,

$$\sigma_y \sigma_z + \sigma_z \sigma_y = 0 .$$

The relativistic invariancy of the complete logic set can be demonstrated if the temporal aspect is taken into account. In the same way as σ_x, σ_y, σ_z are associated with space coordinates, the identity σ_0 can be associated with the time coordinate σ_t.
The spin operator σ can then be expressed via the operators of a logic base as follows:

$$R\sigma = T\sigma_t + X\sigma_x + Y\sigma_y + Z\sigma_z$$

where R = (T, X, Y, Z) is a unit space-time vector.
Unitary transformations in spinor space may influence an individual component of a space-time vector, but conserve its pseudo-Euclidian length:

$$\det R\sigma = \begin{vmatrix} T - Z & X + iY \\ X - iY & T + Z \end{vmatrix} = T^2 - X^2 - Y^2 - Z^2.$$

This examination explicitly demonstrates that spinor logic is characterized by being Lorentz invariant: since the determinant is invariant with respect to unitary transformation, the expression $T^2 - X^2 - Y^2 - Z^2$ is unchanged, but that is precisely the invariant relativistic interval.

Under the interpretations given above, the operations in the logic vector space correspond to unitary transformations in the spinor space, which in turn are associated with corresponding Lorentz transformations in space-time:

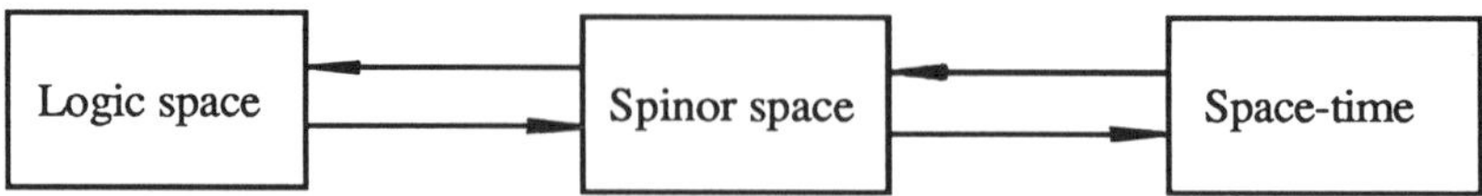

The following scheme presents the set of relativistic spinor logic operators with respect to the nonrelativistic spinor logic subset, and the subsets of operators of positive and negative logics:

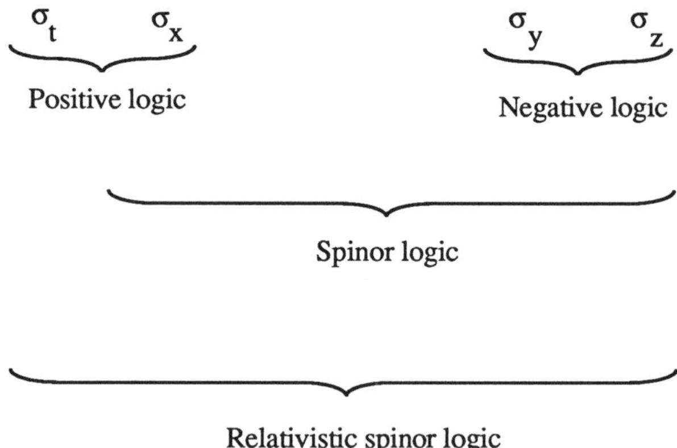

From the physical point of view a positive subset of logic is a system of binary operations that is built around the orthogonal group SO(3) of 3-D space rotations, whereas complete logic, including its positive and negative subsets, is a system of binary operations that is built around the unitary spin group SU(2) in spinor space.

The rotation group SO(3), spin group SU(2) and the Lorentz group of linear transformations, which have found fundamental applications in relativity and quantum physics, can all be conveniently handled by the (2×2) matrix group, making the link with matrix logic quite simple and meaningful.

The common mathematical roots of these three cornerstones of scientific inquiry promise to provide new approaches to the advanced theories of fundamental physics.

In the previous chapter we showed that the problem of the functional completeness of logic can be addressed in algebraic terms as the problem of the existence of a complete matrix base. The proposal to link logic and spin derives additional support from the fact that the elements σ_t, σ_x, σ_y, σ_z of the spin group SU(2) constitute a linear independent matrix base from which any (2×2) matrix can be generated, and conversely, given an operator, we can uniquely decompose it into the elements of the spinor matrix base with corresponding multipliers.

In the relativistic case we replace the 2-component Pauli spinor with the 4-component Dirac spinor:

$$\psi = \begin{pmatrix} \psi_1 \\ \psi_2 \\ \psi_3 \\ \psi_4 \end{pmatrix}; \qquad \psi^* = (\psi_1^* \ \psi_2^* \ \psi_3^* \ \psi_4^*)$$

and the Pauli matrix base σ_0, σ_x, σ_y, σ_z by Dirac spinor matrices:

$$\gamma_1 = \begin{pmatrix} 0 & \sigma_x \\ -\sigma_x & 0 \end{pmatrix} \quad \gamma_2 = \begin{pmatrix} 0 & \sigma_y \\ -\sigma_y & 0 \end{pmatrix} \quad \gamma_3 = \begin{pmatrix} 0 & \sigma_z \\ -\sigma_z & 0 \end{pmatrix} \quad \gamma_4 = \begin{pmatrix} \sigma_0 & 0 \\ 0 & -\sigma_0 \end{pmatrix}.$$

Altogether the Dirac matrices generate 16 different and linearly independent products γ_i, $i = 1, 16$, which can be associated with 16 different logic operators.

$$\begin{aligned}
&\gamma_0 = I, && \gamma_{10} = i\,\gamma_2\,\gamma_4, \\
&\gamma_1,\ \gamma_2,\ \gamma_3,\ \gamma_4,\ \gamma_5 = \gamma_1\,\gamma_2\,\gamma_3\,\gamma_4, && \gamma_{11} = i\,\gamma_3\,\gamma_4, \\
&\gamma_6 = i\,\gamma_1\,\gamma_2, && \gamma_{12} = i\,\gamma_1\,\gamma_2\,\gamma_3, \\
&\gamma_7 = i\,\gamma_2\,\gamma_3, && \gamma_{13} = i\,\gamma_1\,\gamma_2\,\gamma_3, \\
&\gamma_8 = i\,\gamma_3\,\gamma_1, && \gamma_{14} = i\,\gamma_3\,\gamma_1\,\gamma_4, \\
&\gamma_9 = i\,\gamma_1\,\gamma_4, && \gamma_{15} = i\,\gamma_1\,\gamma_2\,\gamma_4.
\end{aligned}$$

The coefficients for γ_i are chosen to satisfy

$$\gamma_i^2 = \gamma_0$$

where $\gamma_0 = I$ is the (4×4) unit matrix.
While the algebra of the Pauli (2×2) spin matrices is the quaternion algebra, relativistic spinor logic is associated with a Clifford algebra or hypercomplex algebra. The matrices $\gamma_1\ \gamma_2\ \gamma_3\ \gamma_4$ generate a matrix base of 16 linearly independent quantities in which any (4×4) matrix can be expanded.
Consequently, to investigate the properties of relativistic spinor logic it would be sufficient to know the behaviour of the 16 independent bilinear products

$$<\psi^* | \gamma_i | \psi> \qquad i = 1, 16.$$

In the domain of relativistic logic it is worth pointing out an important case: when given a fundamental product

$$<\psi_a | \gamma_i | \psi_b><\psi_c | \psi_j | \psi_d>$$

where ψ_a, ψ_b, ψ_c, ψ_d are bispinors, γ_i, γ_j are Dirac matrices, we shall define an equivalent product

$$<\psi_a | \gamma_k | \psi_d><\psi_c | \gamma_t | \psi_b>$$

where γ_k, γ_t are some other Dirac matrices.
This problem is similar to the problem of the recombination of logic vectors within a fundamental product we considered earlier. Consider the following illustration:

$$<\bar{x} | \vee | y><x | \vee | \bar{y}> = <\bar{x} | \leftarrow | \bar{y}><x | \leftarrow | y> = <x | y>$$

$$<\bar{x} | \rightarrow | y><x | \vee | \bar{y}> = <\bar{x} | \downarrow | \bar{y}><x | \leftarrow | y> = <x | 1>$$

$$<\bar{x} | \rightarrow | y><x | \downarrow | \bar{y}> = <\bar{x} | \downarrow | \bar{y}><x | \rightarrow | y> = <1 | y>$$

4.3 CPT-theorem

Spinor logic introduces the methods of symmetry groups into logic, expanding the scope of matrix logic applications. The analysis of symmetry groups is very useful because it allows one to obtain meaningful results on the basis of the study of symmetry properties alone.
An important example is represented by the CPT-theorem, which is considered to be the most general conservation law in physics. Any physical interaction is invariant with respect to the following three fundamental operations of symmetry, executed jointly: Charge inversion,

Parity, and Time reversal. Though single operations of symmetry and some incomplete combined symmetry operations may violate conservation, the complete operation CPT fulfils the most general requirements of relativistic invariancy.

According to the Pauli theorem, relating spin and quantum statistics, any relativistically invariant theory with local interactions ought to be invariant with respect to the CPT-tranformation. If logic is a fundamental phenomenon, then could it fail to comply with the requirements of relativistic covariancy?

We shall attempt to analyse this question with an analogue of the CPT-theorem which can be formulated in matrix logic, both positive and negative. We shall illustrate it on a positive subset of logic only. In order to achieve this goal, we must accommodate the fundamental operations of symmetry - Charge inversion, Parity and Time reversal - in terms of matrix logic operations.

Charge inversion. The analogue of charge inversion in logic is accommodated by Boolean complementation of the reference truth-values: 0 is complemented by 1, and 1 is complemented by 0. The complementation induces on the set of logic operators Ω^+ the following transformations:

	L	CL	
	$\begin{pmatrix} 1 & 0 \\ 0 & 1 \end{pmatrix}$	$\begin{pmatrix} 1 & 0 \\ 0 & 1 \end{pmatrix}$	
	$\begin{pmatrix} 0 & 1 \\ 1 & 0 \end{pmatrix}$	$\begin{pmatrix} 0 & 1 \\ 1 & 0 \end{pmatrix}$	
I	$\begin{pmatrix} 1 & 1 \\ 1 & 1 \end{pmatrix}$	$\begin{pmatrix} 1 & 1 \\ 1 & 1 \end{pmatrix}$	CL = L
	$\begin{pmatrix} 0 & 0 \\ 0 & 0 \end{pmatrix}$	$\begin{pmatrix} 0 & 0 \\ 0 & 0 \end{pmatrix}$	

II

$\begin{pmatrix}0&0\\0&1\end{pmatrix}$	$\begin{pmatrix}1&0\\0&0\end{pmatrix}$
$\begin{pmatrix}0&1\\1&1\end{pmatrix}$	$\begin{pmatrix}1&1\\1&0\end{pmatrix}$
$\begin{pmatrix}1&1\\1&0\end{pmatrix}$	$\begin{pmatrix}0&1\\1&1\end{pmatrix}$
$\begin{pmatrix}1&0\\0&0\end{pmatrix}$	$\begin{pmatrix}0&0\\0&1\end{pmatrix}$

$CL = \overset{\sqcap}{L}$

III

$\begin{pmatrix}1&1\\0&1\end{pmatrix}$	$\begin{pmatrix}1&0\\1&1\end{pmatrix}$
$\begin{pmatrix}1&0\\1&1\end{pmatrix}$	$\begin{pmatrix}1&1\\0&1\end{pmatrix}$
$\begin{pmatrix}0&0\\1&0\end{pmatrix}$	$\begin{pmatrix}0&1\\0&0\end{pmatrix}$
$\begin{pmatrix}0&1\\0&0\end{pmatrix}$	$\begin{pmatrix}0&0\\1&0\end{pmatrix}$

$CL = \tilde{L}$

IV			$CL = \overline{L}$
	$\begin{pmatrix} 0 & 0 \\ 1 & 1 \end{pmatrix}$	$\begin{pmatrix} 1 & 1 \\ 0 & 0 \end{pmatrix}$	
	$\begin{pmatrix} 1 & 1 \\ 0 & 0 \end{pmatrix}$	$\begin{pmatrix} 0 & 0 \\ 1 & 1 \end{pmatrix}$	
	$\begin{pmatrix} 0 & 1 \\ 0 & 1 \end{pmatrix}$	$\begin{pmatrix} 1 & 0 \\ 1 & 0 \end{pmatrix}$	
	$\begin{pmatrix} 1 & 0 \\ 1 & 0 \end{pmatrix}$	$\begin{pmatrix} 0 & 1 \\ 0 & 1 \end{pmatrix}$	

Examination of the operators obtained shows that the set Ω^+ is divided into 4 quadruplets of operators, each reacting differently to the complementation of the reference truth-values. In the first quadruplet, complementation is an identity operation; in the second, it is reflection; in the third, transposition; and in the fourth, matrix complement.

Fortunately, these four different responses to Boolean complementation can be uniformly described by a single operation, the orthogonal transformation θ of logic operators:

$$CL = \theta L = \rceil L \rceil.$$

<u>Parity operation.</u> Transformation of a logic operator under the parity operation can be equally well described either by transposition or by converse transposition (reflection). Because transposition in matrix logic describes antecedent-consequent recombination, it must be reserved for the operation of time reversal. Consequently, we shall use reflection $\overset{\sqcap}{L}$ as the parity operation. Applying it to the set Ω^+, after Boolean complementation has been carried out, we obtain two quadruplets of operators, which remain invariant under reflection, while two others are modified accordingly.

<u>Time reversal.</u> Finally, the operation of time reversal recombines cause and effect by the transposition operation. As a result the first quadruplet is again unchanged, the second is also unchanged, but the third and fourth now undergo transposition:

L	θL	$\theta \overline{L}$	$\theta \widetilde{\overline{L}}$
↔	↔	↔	↔
⌉	⌉	⌉	⌉
$\hat{1}$	$\hat{1}$	$\hat{1}$	$\hat{1}$
$\hat{0}$	$\hat{0}$	$\hat{0}$	$\hat{0}$
∧	↑	∧	∧
∨	↓	∨	∨
↓	∨	↓	↓
↑	∧	↑	↑
→	←	←	→
←	→	→	←
\	/	/	\
/	\	\	/
⊥	T	⊣	⊥
T	⊥	⊢	T
⊣	⊢	⊥	⊣
⊢	⊣	T	⊢

Comparing the set of logic operators with the operators obtained by the complete symmetry operation CPT, we observe that they have returned to their initial form:

$$\theta\tilde{\bar{L}} = L\,.$$

Within the framework of logic the operation of symmetry is of special significance for implication. Implication would be conserved under any transformation in which cause and effect recombination is accompanied by a change-over from true to false and vice versa.
In physics symmetry has established itself as one of the basic concepts. Though symmetry relations are deeply present at the very foundation of logic, they have never been considered as occupying a focal point there. The matrix formulation of logic places this problem in a new perspective.
The isomorphism of the symmetry groups in logic and physics, in spite of the different nature of the group elements, could allow one to study these processes either from a physical viewpoint or from the point of view of logic.
In a more dramatic instance, an opportunity may arise when it would become possible to introduce logic operators into physical description and physical operators into logical description, leading to a sort of logico-physical conservation principle. In such an event, CPT conservation in physics and in logic would be given not just the status of two aspects of reality with reflexive symmetry between them, but would be actually united in a common basis. The relativistic covariance of logic provides an explanation for why mathematically correct results of our intelligence are independent in their outcome of the particular frame of reference.

4.4 Fundamental interactions

Direct links to the spinor formalism and the relativistic invariance of logic are indications that matrix logic can contribute fruitfully to ongoing research into fundamental interactions, and especially to the classification of quantum fields and particles. Furthermore, the application of matrix logic to fundamental physics promises to be mutually beneficial to both fields.
It seems useful to illustrate the way in which matrix logic can find applications in this vast domain.

Logic model of the atom

From the logic point of view an atom can be interpreted as a codeword, where the corresponding Boolean values are assigned to the two states of a nucleon, charged proton and neutral neutron:

$$N := \begin{cases} 1 & - \text{ proton p} \\ 0 & - \text{ neutron n} \end{cases}$$

A wave function of an oriented nucleon duplet is a perfect match to a logic vector, with

$$\begin{pmatrix} n \\ p \end{pmatrix} \text{ equal to } \begin{pmatrix} 0 \\ 1 \end{pmatrix}$$

and

$$\begin{pmatrix} n \\ p \end{pmatrix} \text{ equal to } \begin{pmatrix} 0 \\ 1 \end{pmatrix}.$$

Under these interpretations, the nucleus of such an atom as helium can be, for example, conveniently described by such logic operators as equivalence or negation:

$$\leftrightarrow = \begin{pmatrix} 1 & 0 \\ 0 & 1 \end{pmatrix} \quad , \quad \urcorner = \begin{pmatrix} 0 & 1 \\ 1 & 0 \end{pmatrix}.$$

Generally, in order to define a (2×2) logic operator, it is necessary to be able to specify all 4 of its components. For an aggregate of 4 particles this specification may employ either permitted spin variations or distribution of charges. If spinor logic is employed, the logic operator could be accommodated by two nucleon duplets (deutrons), such as the helium nuclei, with 4 possible spin orientations. A spinor specification of the additive logic base { $\wedge$, /, \, $\uparrow$ } would require a nucleus with 16 nucleons.

Among stable atoms this implies an unique choice, oxygen $_8O^{16}$. Such a model is naturally based on a specification of the resulting spin of the 4 nucleon compound, and not on a specification of charges:

$$\underset{\leftarrow}{\begin{pmatrix} 0 & 0 \\ 0 & 1 \end{pmatrix}} \quad \underset{\downarrow}{\begin{pmatrix} 0 & 1 \\ 0 & 0 \end{pmatrix}} \quad \underset{\uparrow}{\begin{pmatrix} 0 & 0 \\ 1 & 0 \end{pmatrix}} \quad \underset{\rightarrow}{\begin{pmatrix} 1 & 0 \\ 0 & 0 \end{pmatrix}}.$$

In the additive logic base there are altogether 4 "charged" matrix elements (units) and 12 "neutral" matrix elements (zeros), whereas in isotopically invariant nuclei $_8O^{16}$ the number of charged protons and neutral neutron particles is equal. Thus, the use of spinor logic is the only option open in this case.

On the other hand, the multiplicative logic base { ⌉, ∧, ∨ }:

$$\begin{pmatrix} 0 & 1 \\ 1 & 0 \end{pmatrix} \quad \begin{pmatrix} 0 & 0 \\ 0 & 1 \end{pmatrix} \quad \begin{pmatrix} 0 & 1 \\ 1 & 1 \end{pmatrix}$$

is perfectly suited to being encoded by such isotopically invariant nuclei as even-even carbon nuclei $_6C^{12}$. The 6 protons and 6 neutrons of $_6C^{12}$ could encode the three multiplicative generators of logic.

The adaptive logic base { ↔, ⌉, ∧, /, \, ↑, $\hat{0}$ } brings us back to $_8O^{16}$, since altogether it contains 8 charged and 8 neutral elements.

Strong ternary switch

Change-over from proton to neutron and vice versa, occurring both in the strong and weak interactions, on one hand and a logic representation of two states of a nucleon in the form of logic vectors on the other allow for interpretation of these physical processes in terms of logic switch operations. In the strong switch scheme presented below, the arrows indicate absorption and emission routes.

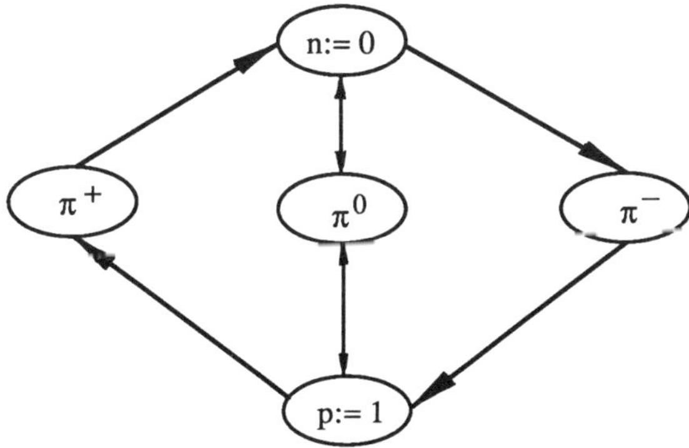

Ternary strong switch

Change-over proton-neutron is provided by interactions with the meson triplet $\{\pi^+ \ \pi^0 \ \pi^-\}$. A proton accepting π^- or emitting π^+ is translated into a neutron; a neutron accepting π^+ or emitting π^- is translated into a proton. Absorption or emission of neutral π^0 does not modify the state of a nucleon. Presenting a proton and neutron in the form of corresponding logic vectors, (0,1) and (1,0) respectively, interaction with π^0 can be conveniently described by the identity operation:

n:

$$(1,0)\begin{pmatrix}1 & 0\\ 0 & 1\end{pmatrix} = (1,0) \qquad \text{– composition equation}$$

$$\begin{pmatrix}1 & 0\\ 0 & 1\end{pmatrix}\begin{pmatrix}1\\ 0\end{pmatrix} = \begin{pmatrix}1\\ 0\end{pmatrix} \qquad \text{– decomposition equation}$$

p:

$$(0,1)\begin{pmatrix}1 & 0\\ 0 & 1\end{pmatrix} = (0,1) \qquad \text{– composition equation}$$

$$\begin{pmatrix}1 & 0\\ 0 & 1\end{pmatrix}\begin{pmatrix}0\\ 1\end{pmatrix} = \begin{pmatrix}0\\ 1\end{pmatrix} \qquad \text{– decomposition equation}$$

Interactions with $\pi^+$and π^- represent oriented negations, and operators describing π^+ and π^- can be extracted from the negation operator:

$$\begin{pmatrix}0 & 1\\ 1 & 0\end{pmatrix} = \begin{pmatrix}0 & 1\\ 0 & 0\end{pmatrix} + \begin{pmatrix}0 & 0\\ 1 & 0\end{pmatrix},$$

where the converse inhibition operator corresponds to π^+, and the inhibition operator corresponds to π^-. The equations of composition and decomposition will thus be:

n:

$$(1,0)\begin{pmatrix}0 & 1\\ 0 & 0\end{pmatrix} = (0,1) \qquad \text{– composition equation}$$

$$\begin{pmatrix}0 & 0\\ 1 & 0\end{pmatrix}\begin{pmatrix}1\\ 0\end{pmatrix} = \begin{pmatrix}0\\ 1\end{pmatrix} \qquad \text{– decomposition equation}$$

p:

$$(0,1)\begin{pmatrix}0 & 0\\ 1 & 0\end{pmatrix} = (1,0) \qquad \text{– composition equation}$$

$$\begin{pmatrix}0 & 1\\ 0 & 0\end{pmatrix}\begin{pmatrix}0\\ 1\end{pmatrix} = \begin{pmatrix}1\\ 0\end{pmatrix} \qquad \text{– decomposition equation}$$

Weak binary switch

The weak switching of nucleons is strictly binary. It involves the following fundamental reactions:

$$n = p + e^{-} + \tilde{\nu}$$

$$p = n + e^{+} + \nu$$

$$p + e^{-} = n + \nu$$

$$p + \tilde{\nu} = n + e^{+}$$

where e^{-} and e^{+} are an electron and a positron, and ν and $\tilde{\nu}$ a neutrino and an anti-neutrino. The weak switch can thus be schematically presented as follows:

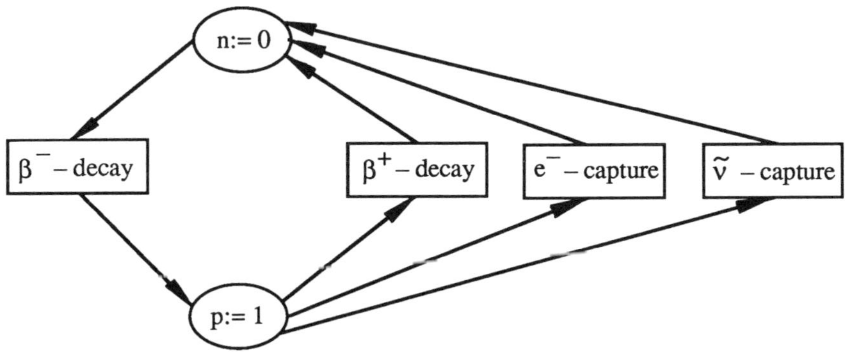

Weak binary switch

Though strictly binary in comparison to the strong nucleon switch, the weak switch is asymmetrically disbalanced, reflecting a fundamental shift of the accumulation of neutrinos.
A logic description of the switch can be achieved with the disjunction operator and its inverse for the decaying section, yielding the following decomposition equations:

$$\beta^{-}: \quad \begin{pmatrix} 0 & 1 \\ 1 & 1 \end{pmatrix} \begin{pmatrix} 1 \\ 0 \end{pmatrix} = \begin{pmatrix} 0 \\ 1 \end{pmatrix}$$

$$\beta^{+}: \quad \begin{pmatrix} -1 & 1 \\ 1 & 0 \end{pmatrix} \begin{pmatrix} 0 \\ 1 \end{pmatrix} = \begin{pmatrix} 1 \\ 0 \end{pmatrix}$$

with the identity operation being available:

$$\begin{pmatrix} 0 & 1 \\ 1 & 1 \end{pmatrix} \begin{pmatrix} -1 & 1 \\ 1 & 0 \end{pmatrix} = I.$$

The switching occurring in the e^- - capture and $\tilde{\nu}$ - capture can be accommodated by composition equations:

$$\tilde{\nu}: \quad (0 \quad 1) \begin{pmatrix} 1 & 0 \\ 1 & 0 \end{pmatrix} = (1 \quad 0)$$

$$e^{-}: \quad (0 \quad 1) \begin{pmatrix} -1 & 0 \\ 1 & 0 \end{pmatrix} = (1 \quad 0)\,,$$

with no identity operation being available and an imbalance between two processes tending towards e^- - capture switching:

$$\begin{pmatrix} -1 & 0 \\ 1 & 0 \end{pmatrix} \begin{pmatrix} 1 & 0 \\ 1 & 0 \end{pmatrix} = \begin{pmatrix} -1 & 0 \\ 1 & 0 \end{pmatrix}.$$

Annihilating interactions

Particle - anti-particle interaction results in annihilation. As we can recall from the previous chapter, the interaction of two non-zero logic operators can also result in the empty operator. This indicates the possibility of expressing a particle - anti-particle annihilation by employing annihilating logic operators. Consider an annihilation of electron and positron in which two charge-free gamma quanta are born:

$$e^- + e^+ = 2\gamma.$$

In order to provide a matrix logic description of this process, recall that in the set Ω^+, we have an octet of annihilating operators:

$$\{ \wedge, /, \backslash, \uparrow \} \quad \text{and} \quad \{\perp, \top, \dashv, \vdash \}$$

Of these, only operators with even "weight" (the number of the 1's must be equal to the number of 0's) are suitable for the description of a pair of charged particle and anti-particle, since charge conjugation must not change the "weight" of the operator but only its structure. This leaves us with $\{\perp, \top, \dashv, \vdash\}$ as the adequate set. Let us associate electron e^- with the operator $\perp$. The change-over from e^- to anti-particle e^+ is naturally attained by complementation

$$\perp^* = \overline{\perp} = \theta \perp = \top$$

and transposition

$$\tilde{\top} = \vdash .$$

Interaction of e^-, represented by $\perp$, and e^+, represented by $\vdash$, is an annihilating interaction

$$\vdash \perp = \hat{0} \ .$$

Its conjugate transpose is also annihilating:

$$\dashv \top = \hat{0} \ .$$

The corresponding bilinear products will thus be written not simply as $\langle e^- | \hat{0} | e^+ \rangle$ but as

$$\langle e^+ | \vdash \perp | e^- \rangle \qquad \text{or} \qquad \langle e^- | \dashv \top | e^+ \rangle$$

where $\langle e^+|$ and $|e^-\rangle$ are two dual vectors.

Spin propagator

Consider a fermion lattice of interacting spins:

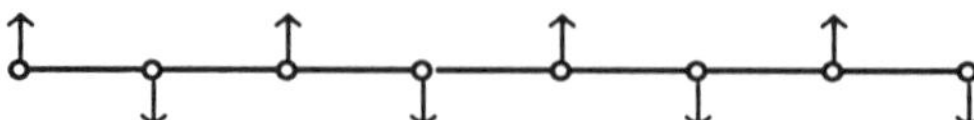

In the formalism of matrix logic it can be described by a multiplication of inner products:

$$<x_1 | y_1><x_2 | y_2><x_3 | y_3> \ldots <x_n | y_n> .$$

Spin inversion at the beginning of the lattice can be carried through the lattice as a spinor wave. Propagation of spinor wave transfers spin inversion from x_1 to y_n, with "communication" between the spins complying with the Pauli exclusion principle.

Transfer of spin inversion from the first position to the last position can be described as orthogonal transformation

$$<\bar{x}_1 | L | y_n> = <x_1 | \theta L | \bar{y}_n>$$

where

$$L = |y_1><x_2 | y_2><x_3 > y_4> \ldots <x_n|$$

and θL is a logic operator computable by the lattice.

Schematically this process can be described as a spinor wave propagating from one end of the lattice to the other:

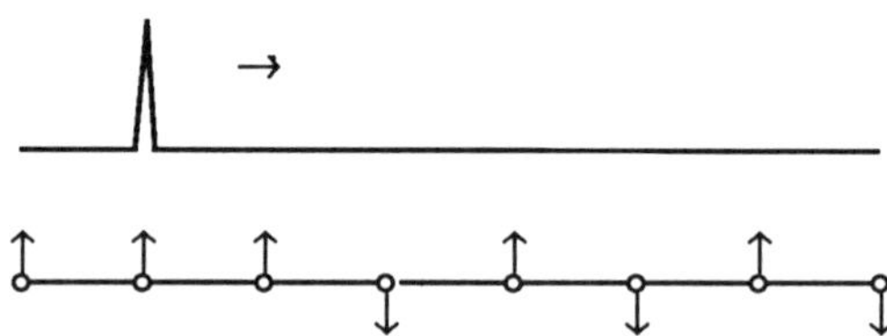

The process is ignited by a single spin inversion at the first position. If perturbation of spin occurs in the centre of the lattice, the latter can be modified by two spinor waves, spreading in opposite directions out from the centre. Such a process is triggered by inversions of the pair of spins in the centre of the lattice

$$\ldots | \bar{y}_i><\bar{x}_j| \ldots$$

which is equivalent to the orthogonal transformation of a corresponding kern operator. The result of out-from-the-centre propagation is accommodated by the following matrix logic equation:

$$<x_1 \,|\, L_{bra} \,|\, \bar{y}_i><\bar{x}_j \,|\, L_{ket} \,|\, y_n> = <\bar{x}_1 \,|\, \theta L_{bra} \,|\, y_i><x_j \,|\, \theta L_{ket} \,|\, \bar{y}_n>$$

where operators of the bra-branch and ket-branch of the lattice are respectively:

$$L_{bra} = |y_1><x_2 \,|\, y_2><x_3 \,|\, y_3> \dots <x_i|$$

$$L_{ket} = |y_j><x_{j+1} \,|\, y_{j+1}><x_{j+2} \,|\, y_{j+2}> \dots <x_n|$$

Spin propagators can be interpreted simultaneously as a physical perturbation process and as a logic process.

Space-time diagrams

In the space-time diagram method, particles are represented by dots connected by oriented paths; continuous paths depict the space-time map of a particle, the dashed lines corresponding to their interactions.

Since the time axis is directed from the bottom up, the paths directed towards the bottom are associated with the anti-particles. The horizontal axis represents 3-D space. The diagrams below describe several cases of interactions:

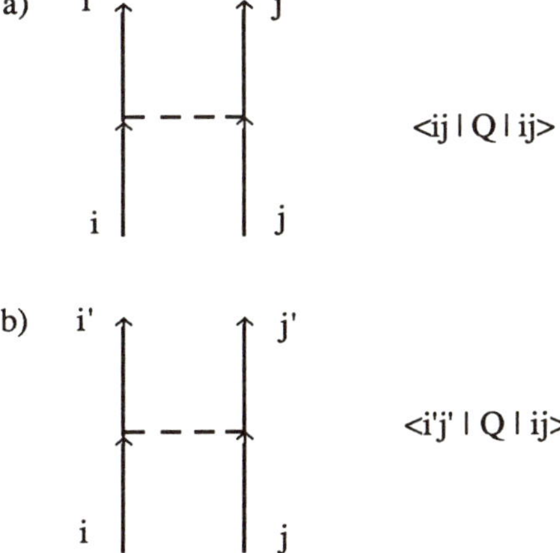

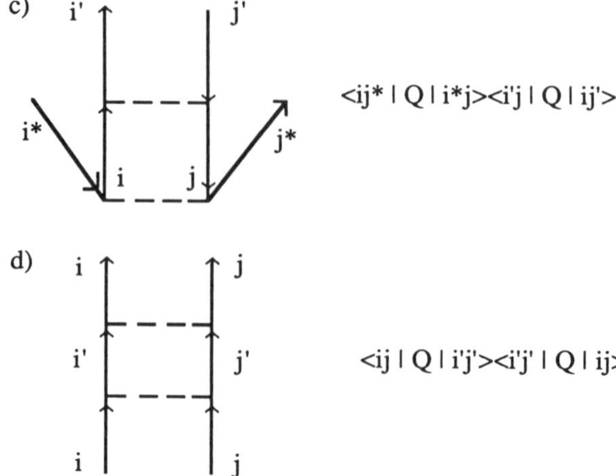

Diagram a) depicts two fermions in the states i and j which as a result of interaction remain unchanged. Diagram b) depicts two interactions in which the initial states are dispersed into the new states i' and j'.

The interaction in diagram c) involves anti-particles. Diagrams c) and d) describe two consecutive interactions. In the latter case, fermions by the first interaction are dispersed from initial states i and j into intermediary states i' and j' , and in the second interaction dispersed into final states, which are here the same as the initial ones.

The space-time diagrams have an exact correspondence with the mathematical expressions of quantum field theory. The diagrams a) and b) with a single interaction are associated with the following matrix elements respectively

$$\langle ij \mid Q \mid ij \rangle \quad \text{and} \quad \langle i'j' \mid Q \mid ij \rangle$$

where i,j,i',j' are eigenstates of the momentum operator, and Q is the quantum-mechanical operator.

Matrix elements associated with diagram d) are written as a normal product:

$$\langle ij \mid Q \mid i'j' \rangle \langle i'j' \mid Q \mid ij \rangle,$$

which describes two consecutive acts of interaction.

The normal products provide radiative corrections to the energy of a ground state E_0 . A correction of order r, $E^{(r)}$, is associated with a diagram with r dashed lines of interactions, whereas the number of continuous paths is even and can vary from 2 to 2r. The normal products will contain r multipliers.

A parallel can be drawn between the normal products of quantum field theory and the fundamental product reductions we derived in matrix logic. In the same way as the complete diagram of fundamental interactions involves multiplication of normal products, a compound logic expression is also reducible to a fundamental product of the same multiplicative structure. According to the Reduction Theorem of matrix logic any logic expression can be reduced to a product of bilinear products. Consequently, it might be possible to put in correspondence to a given diagram, a corresponding matrix logic expression and vice versa.
Compare for example the identity transformation in the diagram d)

$$\langle ij | Q | i'j' \rangle \langle i'j' | Q | ij \rangle$$

with a decomposition of the equivalence product:

$$\langle x | y \rangle = \langle x | \vee | \bar{y} \rangle \langle \bar{x} | \vee | y \rangle$$

The multiplicative structure of both expressions is identical.

In view of the covariance of logic operators under the complete operation of symmetry the idea of relating the fundamental reductions of matrix logic and the diagram method takes on special importance.
A significant feature of the application of matrix logic to particle physics consists in the prospect of placing at the basis of fundamental interactions a logic operation with universal group properties. If this could be effectively achieved, it would be possible to describe fundamental interactions as logic networks, whereas on the other hand logic processes could be interpreted and perhaps studied with the help of physical methods.

Until recently conventional logic was developing in postulatory fashion, making great use of heuristics, and without being grounded in fundamental principles. This approach cannot reveal the relativistic covariance of the theory, nor does it show the close links with group-theoretic analysis. On the other hand, these links alone are almost sufficient to demonstrate the absolute necessity of such a theory from initial principles.
On the premiss that the isomorphism of logic and fundamental interactions is not coincidental, but reflects fundamental relations, one may seek a synthesis of physics and logic into a unified theory.
Traditionally physics was preoccupied with problems of the study of motion. With the development of fundamental physics interest has shifted towards the study of symmetry. It is possible that there is an even more general approach based on logic. The task of reformulating

physical theory in terms of fundamental logic is laborious but promises to have a unifying effect of major significance.

Spinor implication

The proposal to view logic as the macroanalogue of spin discussed in this chapter allows one to seek the resolution of important logical problems in the context of spin-related phenomena. If indeed the logic structures are isomorphic, as suggested, with spinor logic, then we can hope to detect manifestations of this isomorphism which conventional logic is unable to either predict or explain. In much the same way as macroscopic evidence for quantum effects can be found in physics by setting up proper experimental conditions, we can also arrange ordinary discourse in such a way as to observe evidence of the spinor foundation of logic.
The most persuasive examples in which spin-like behaviour effects can be observed are self-referential discourse reasoning structures. Consider the fundamental paradox formulated by Russell:

$$R \in R = \overline{R \in R} \qquad \text{or} \qquad R \in R = R \notin R$$

where R is a set of all sets which are not subsets of themselves.
In the Russell paradox, the classical rule of inference enforces the logical equivalence of two mutually exclusive propositions:

$$((x \rightarrow \bar{x}) \wedge (\bar{x} \rightarrow x)) \rightarrow (x \leftrightarrow \bar{x})$$

where x denotes $R \in R$.
While in the Russell paradox conventional logic is unable to explain how evaluations that are absolutely correct from the mathematical point of view lead to a contradiction, the spin hypothesis not only seems to provide a plausible explanation, but predicts, even requires, the appearance of such "paradoxes".

To investigate this question within the framework of spinor logic we must recall that for spinor we distinguish two different types of symmetry: mirror symmetry P and rotational symmetry. Under the operation of mirror symmetry a spinor changes into its complement:

$$P\psi = \overline{\psi}$$

while the square P^2 is identity that leaves a spinor intact:

$$P^2\psi = \psi$$

In logical terms this is equivalent to saying that negation of negation is affirmation that leaves a logic expression unchanged. For a logic vector, instead of considering the square of the mirror operation, we can equally well assume that full circle 2π-revolution has taken place. For an ordinary vector space this operation of rotational symmetry is identical with the square of mirror symmetry P^2: both cases represent the identity operation and are therefore indistinguishable. For spinor the relation between mirror and rotational symmetries are fundamentally altered. The full revolution of a spinor is not an identity but an anti-identity:

$$\psi(\alpha + 2\pi) = -\psi(\alpha) ,$$

where α specifies the initial state.
Let us denote this spinor logic operation as circular implication:

$$\psi \looparrowright -\psi \qquad \text{or} \qquad -\psi \looparrowright \psi$$

the domain of which is the balanced set of truth-values:

$$E_3 = \{ -1, 0, 1\}.$$

Circular implication represents a uniquely spinor logic operation, which establishes a dynamic link between the positive and negative subsets of logic that is both natural and fundamental. Spinor implication is an immensely useful operation which tells us precisely how the representation of a logic vector is affected if the operation of rotational symmetry is executed. Associating the antecedent with a spinor $\psi(\alpha)$ in the positive domain and the circular consequent with a spinor $\psi(\alpha + 2\pi)$, we shall run into no inconsistencies if and only if we define the circular consequent in the negative domain, where it is equal to $-\psi(\alpha)$. In so doing we not only link constructively positive and negative logics but also draw a line between that which is true or false by assignment, and that which is true or false by being actually realized. The classical way of truth valuation is at fault in the Russell paradox because it is applied to a circularly closed statement. Because circular closure is not taken into account, we erroneously infer the conventional biconditional:

$$x \leftrightarrow \bar{x} .$$

However, when circular spinor implication is applied x is not equal to $\bar{x}$ but becomes $-x$ upon a full circle of logic evaluation:

$$x \circ\!\!\rightarrow -x \qquad \text{and} \qquad -x \circ\!\!\rightarrow x\,.$$

Consequently, for spinor implication the biconditional cannot be written in classical form

$$((x \rightarrow y) \wedge (y \rightarrow x)) \rightarrow (x \leftrightarrow y)$$

but becomes

$$((x \circ\!\!\rightarrow y) \wedge (y \circ\!\!\rightarrow x)) \rightarrow (x \leftarrow\!\circ\!\rightarrow x)\,.$$

This is a direct consequence of the fact that identity or equivalence for a spinor is not $I(2\pi)$ but rather double-group identity $I^2(2\pi) = I(4\pi) = I_{spinor}$.

If the proper truth valuation of a closed self-referential proposition requires the use of the spinor biconditional, then the Russell paradox can be seen as an unfortunate attempt to apply conventional identity where spinor identity ought to be used. The truth valuation process is not carried out completely to satisfy the double-group identity requirement but it is interrupted in the middle of valuation.

This is precisely what the conventional rule of inference ignores and, consequently, it is unable to explain where and why failure occurs in the Russell paradox.

The use of the proper identity operation not only avoids contradiction, making the Russell paradox obsolete in spinor logic, but also shows that the latter is an absolutely correct circular tautology of matrix spinor logic.

Indeed, whereas in strictly positive logic

$$\langle x \mid \rightarrow \mid \bar{x}\rangle\langle \bar{x} \mid \rightarrow \mid x\rangle \equiv 0$$

is a contradiction, in complete logic employing spinor implication we obtain a tautology:

$$\langle x \mid \circ\!\!\rightarrow \mid -x\rangle\langle -x \mid \circ\!\!\rightarrow \mid x\rangle \equiv 1.$$

The introduction of negative logic and spinor double-group identity breaks new ground in a subject already considered to be classic, the Gödel Incompleteness Theorem. The latter can be seen as a direct consequence of the spinor foundation of logic. In the framework of the interpretation of logic as the macro-analogue of spin the problem of provability is replaced by the problem of measurability, subject to the limitations of the quantum uncertainty principle.

Because the results of logic measurements are associated with the quadratic form of a spinor

$$\psi^2 = \begin{cases} 0 & \text{– false} \\ 1 & \text{– true} \\ -1 & \text{– negative,} \end{cases}$$

taking the square root of ψ^2 we can extract the complete set of truth-values of spinor logic:

$$\psi = \begin{cases} 0 & \text{– false} \\ 1 & \text{– true} \\ -1 & \text{– negative} \\ i & \text{– imaginary} \end{cases}$$

Thus every positive truth-value has its negative counter image

$$-1^2 = 1$$

and every negative truth-value has its imaginary counter image

$$i^2 = -1.$$

Tensor invariants

Throughout this book we have mostly been preoccupied with the multiplicative aspect of matrix logic. However, important features of logic become apparent in the joint approach which explores both the multiplicative and the additive properties of the formalism.
In the inner product representation, which generally contains the complement operators, these can be eliminated if we employ an additive decomposition of the corresponding logic operators. Take a simple example of the inner product representation of the implication function:

$$\langle x \mid \rightarrow \mid y\rangle = \overline{\langle x \mid 1\rangle\langle 0 \mid y\rangle}$$

Decomposing the implication operator into the corresponding matrix sum we can express the implication function as follows:

$$\langle x \mid \rightarrow \mid y\rangle = \langle x \mid 0\rangle\langle 0 \mid y\rangle + \langle x \mid 0\rangle\langle 1 \mid y\rangle + \langle x \mid 1\rangle\langle 1 \mid y\rangle.$$

Thus, in the general case, a matrix logic expression can be presented in the form of a multilinear function, that is a logic sum, containing in every term a certain number of inner products:

$$\alpha = \sum_i \langle x_1^i \mid x_2^i\rangle \ldots \langle x_m^i \mid x_n^i\rangle$$

where $\langle x_1^i|$, $|x_2^i\rangle$, $\langle x_m^i|$, $|x_n^i\rangle$ are variable or constant logic vectors.

It may be apparent that multilinear logic functions expressed with the help of the addition and multiplication of logic vectors are of the same structure as a vector representation of a tensor in a particular basis system. As we know, it is always possible to obtain a tensor by addition of vector products.

An important property of a tensor is that, similar to a vector, it represents a covariant quantity. While components of a tensor may vary in different reference systems, with the passage from one system to another, they are transformed according to certain rules, so that a tensor contains the same information and represents the same quantity.

In revealing the tensor nature of logic the computational reform accomplished in the matrix formulation of logic achieves its utmost significance. According to the covariance principle any correctly formulated physical law must be the same in all reference systems. The construction of tensor invariants in logic enable us to approach the problem of the invariance of the laws of logic from the same viewpoint, thus bringing logic fully into the realm of the exact sciences.

The construction of tensor invariants in logic can be done in a number of ways. We shall consider here only the most important points of interpretation of logic as a tensor.

First of all it is apparent that, because logic vectors are the basis vectors, we can obtain the tensors of tautology and contradiction, which will be invariant in all reference systems, that is, under any interpretation.

A change of interpretation thus can be viewed as a passage from one basis to another. In a given basis the set of 81 operators of complete logic $\Omega^+ \cap \Omega^-$ generates the set of 81 numbers, which can be associated with the tensor of rank 4. Convoluting the tensor of complete logic we descend to the domain of generalized logic functions defined in the set $E_3 = \{-1, 0, 1\}$:

Equivalence

$$\begin{bmatrix} 1 & 0 & 0 \\ 0 & 1 & -1 \\ 0 & -1 & 1 \end{bmatrix}$$

Negation

$$\begin{bmatrix} 0 & 1 & -1 \\ 1 & 0 & 0 \\ -1 & 0 & 0 \end{bmatrix}$$

Conjunction

$$\begin{bmatrix} -0 & 0 & 0 \\ 0 & 1 & -1 \\ 0 & -1 & 1 \end{bmatrix}$$

Disjunction

$$\begin{bmatrix} 0 & 1 & -1 \\ 1 & 1 & -1 \\ -1 & -1 & 1 \end{bmatrix}$$

Implication

$$\begin{bmatrix} 1 & 1 & -1 \\ 0 & 1 & -1 \\ 0 & -1 & 1 \end{bmatrix}$$

Converse implication

$$\begin{bmatrix} 1 & 0 & 0 \\ 1 & 1 & -1 \\ -1 & -1 & 1 \end{bmatrix}$$

Inhibition

$$\begin{bmatrix} 0 & 0 & 0 \\ 1 & 0 & 0 \\ -1 & 0 & 0 \end{bmatrix}$$

Converse inhibition

$$\begin{bmatrix} 0 & 1 & -1 \\ 0 & 0 & 0 \\ 0 & 0 & 0 \end{bmatrix}$$

Nor

$$\begin{bmatrix} 1 & 0 & 0 \\ 0 & 0 & 0 \\ 0 & 0 & 0 \end{bmatrix}$$

Nand

$$\begin{bmatrix} 1 & 1 & -1 \\ 1 & 1 & 0 \\ -1 & 0 & 0 \end{bmatrix}$$

Converse 1-negation

$$\begin{bmatrix} 0 & 0 & 0 \\ 1 & 1 & -1 \\ -1 & -1 & 1 \end{bmatrix}$$

1-negation

$$\begin{bmatrix} 0 & 1 & -1 \\ 0 & 1 & -1 \\ 0 & -1 & 1 \end{bmatrix}$$

Converse 0-negation

$$\begin{bmatrix} 1 & 1 & -1 \\ 0 & 0 & 0 \\ 0 & 0 & 0 \end{bmatrix}$$

0-negation

$$\begin{bmatrix} 1 & 0 & 0 \\ 1 & 0 & 0 \\ -1 & 0 & 0 \end{bmatrix}$$

Metatautology

$$\begin{bmatrix} 1 & 1 & -1 \\ 1 & 1 & -1 \\ -1 & -1 & 1 \end{bmatrix}$$

Empty constant

$$\begin{bmatrix} 0 & 0 & 0 \\ 0 & 0 & 0 \\ 0 & 0 & 0 \end{bmatrix}$$

Tensors of the generalized logic functions as tensors of rank 2 are defined by the set of 9 components. The affimation tensor of complete logic is of rank 3 containing 27 components, which is precisely, as we have shown earlier, the number of different ways to obtain the identity operator.

The following chart presents specifications for the tensors of complete logic:

Tensor rank	Number of components	Logic quantity
0	$3^0 = 1$	Logic scalar
1	$3^1 = 3$	Generalized logic vector
2	$3^2 = 9$	Generalized logic function
3	$3^3 = 27$	Generalized affirmation
4	$3^4 = 81$	Logic $\Omega^+ \cap \Omega^-$

For the projection of logic in 2-D space we obtain the following quantities:

Tensor rank	Number of components	Logic quantity
0	$2^0 = 1$	Boolean scalar
1	$2^1 = 2$	Logic vector
2	$2^2 = 4$	Logic affinor
3	$2^3 = 8$	Complement operator
4	$2^4 = 16$	Logic Ω^+
5	$2^5 = 32$	Generalized negation operator

As a tensor of rank 4 thc tensor of complete logic can be derived as a tensor outer product of two tensors of rank 2.
With the 81 components of the tensor of complete logic we can construct the invariant

$$R = \langle e_i \,|\, L^i_j \,|\, e^j \rangle = 0, \quad i,j = 1, 81,$$

where L^i_j are one-time covariant and one-time contravariant logic operators, and $\langle e_i|$, $|e^j\rangle$ are covariant and contravariant basis logic vectors respectively.

Quantity R is invariant in all basis systems and can be broken into three polynomial terms:

$$R = \sum_{i=1}^{27} r_i + \sum_{j=1}^{27} r_j + \sum_{k=1}^{27} r_k = 0,$$

where r_i, r_j, r_k are either positive or negative or null components of the tensor.
While for the strictly positive subset of logic Ω^+ the invariant R cannot be constructed, in complete logic $\Omega^+ \cap \Omega^-$ execution of the global logic transformation is characterized by anti-symmetric closure:

$$R = -R,$$

and is invariant in any basis system.
The fact that complete logic is the tensor of rank 4 provides a compelling reason to consider a possible linkage between logic and space anisotropy not only in the mathematical but possibly also in the physical sense of this word.
If this is indeed the case then deviations of the invariant R from its zero ground value can be explained in terms of space deformation, suggesting an interesting connection of logic with general relativity.
The tensor approach allows us to clarify two important questions, the first concerning the evenness of matrix logic and the second related to its classical scalar limit. So far both questions have been treated in a postulatory manner. We have simply stated that Boolean logic, which operates over scalar variables, must be the scalar limit of matrix logic, which operates over logic vectors. But with the use of tensors we are now able to obtain the classical scalar limit of logic by rigorous mathematical derivation. Indeed, convolution of a tensor of rank n produces a tensor of rank n–2, that is reduces the rank of a tensor by 2 as a minimum. Therefore, convoluting any tensor of odd rank, we will eventually end up with a vector. But

convolution of a tensor of even rank will reduce a tensor to a tensor of rank 0, that is, a scalar. Convolution of the tensor of complete logic reduces the rank from 4 to 2, giving us a tensor of a generalized logic function. The next convolution of the tensor of rank 2 produces a Boolean scalar, which is invariant in all reference systems. As is now becoming apparent, the evenness of matrix logic can be explained by the fact that reduction to scalar logic is possible if and only if the operation of convolution is applied to a tensor of even rank.

CHAPTER 5

APPLICATION TO NEW GENERATION COMPUTING

5.1 Multiplicative gates

The development of the computer is closely associated with Boolean logic, which forms a cornerstone of computer system design and operations. In current practice the basic funtions of a computer - numerical computation, logic operations and memory - are all accomplished by devices that have two stable staes, one of the states representing 1, the other state representing 0, thus allowing implementation of the binary number system and Boolean algebra.

Many problems in computer design and testing, artificial intellignce and supercomputing are expressed as a sequence of operations on Boolean functions. With the binary system a computer can evaluate the truth value of logic expressions by making use of three logic gates, NOT, AND and OR. More elaborate logic expressions can be constructed from the three basic functions, and so can arithmetic operations such as addition mod 2.
Unfortunately, many of the basic problems, such as verifying whether there exists any assignment of input variables such that a given Boolean expressions evaluates to true, or whether two Boolean expressions denote the same function, may require an amount of computer power or time that grows exponentially with the size of the problem. The equivalent tranformations of Boolean algebra do permit a decrease in these exponential characteristics, yet the conventional representations suffer from noticeable drawbacks, and in the worst case all known approaches perform as poorly as the naive approach of representing functions by their truth tables. In software terms, most programs that process a sequence of operations on Boolean functions have rather erratic behaviour. They proceed at a reasonable pace, but then suddenly collapse, either running out of storage or failing to complete an operation in a reasonable amount of time.

The above-mentioned drawbacks of conventional computing are caused by the additive approach which requires representations of exponential size which often make them quite impractical. In a general purpose computer, both the simplest and the most sophisticated programs are executed on additive processors, the adders making use of the basic operation addition mod 2. The main reason why contemporary computing has this uniform foundation is that Boolean logic is most suitable for the binary operation addtion mod 2, and the reduction of

arbitrary problems to a sequence of additions in the binary number system forms the essence of the diverse capabilities of the present-day computational system, the general-purpose computer.
Addition, actually, is not the most effective operation. In comparison with such an operation as multiplication it can fail in speed, number of computing steps and, consequently, in reliability. However, by the nature of Boolean logic, it seems rather unnatrual to think of a processor which can employ multiplication as a basic operation for general-purpose computations. Within the scope of Boolean logic, in order to implement multiplicative procedures the latter must be reduced to a sequence of additions and not the other way around. It is inherent in the logic of present-day computers that multiplication operations must be translated into a sequence of elementary operations of addtion mod 2.
Clearly, the matrix formulation of logic introducs a new factor into the situation. As a multiplicative logic in the literal sense of this word, it not only provides a foundation for exploring general-purpose multiplicative gates, but makes this approach most natural and mathematically well-founded.

Employing multiplicative representations we may not only avoid exponential explosions, but also achieve a drastic reduction of computing for a certain subset of problems. The detection of a single zero multiplier in an expression of any length uniquely determines the result of valuation, and thus makes further computations irrelevant and unnecessary.
Consequently, the multiplicative approach must be tremendously effective in giving negative answers, though positive answers could be as hard to compute as in the conventional Boolean computer.
The general conclusion that stems form this discussion can be summarized as follows. Whereas conventional computers are built around an adder that sums up binary numbers, matrix logic processors must be built around multiplicative gates that multiply binary vectors. Thus, the additive machine and the multiplicative machine can be related to each other logarithmically.

Multiplicative processors can find applications for a wide range of problems. Matrix algebra is a powerful technique for solving linear partial equations, control theory equations, signal processing problems, the minimax problem of matrix game theory and many other problems in mathematics, physics, economics etc. Apparently, any problem involving a set of independent multistage computations of the same type can be viewed as a matrix-matrix or matrix-vector multiplication problem.Many important computations such as convolution, Fourier transforms and the finite response filter are special intstances of matrix-vector multiplications. The convolution problem can be represented as an inner product computation which is common to a number of computational routines, such as filtering, pattern matching, correlation, interpolation, polynomial evalutaion, multiplication and division.

The discovery of a link between the matrix formalism and logic makes the above list even more significant. Besides giving a solution to the exponential explosion, it dispels the common misconception that the idea of logical analysis by a computer is applicable only to those cases in which the composition rules involve addition mod2.

Matrix logic algebra forms a foundation for multiplicative processors, just as Boolean logic algebra formed a foundation for conventional additive processors. In the next section a design exercise based on this approach will include specifications of processor organization suitable for multiplicative gates and logic operators transmission elements. These also include designs for hypercube operators linking input and output logics, while simultaneously dealing with memory.

With the formulation of matrix logic, the development of multiplicative processors has become feasible - especially those that implement fixed special-purpose routines. The ultimate goal, however, is effecitve use of general-purpose multiplicative processors that can compete with the flexibility of a general-purpose conventional computer.

The main rewards to be expected from the development of multiplicative general-purpose processors are the possibility of resolving the exponentiality problem, expecially drastic for those tasks which evaluate to 0, and new prospects of introducing a higher degree of parallelism and enlarging the intelligent capabilities of the computer.

Matrix logic can serve to unify the fields of logic programming and artificial intelligence and, possibly, is the missing link between knowledge engineering and parallel computer architecture.

Matrix logic provides explicit clues for designing multiplicative processors which can make use of a high degree of parallelism. Due to superassociativity, a multiplicative processor is capable of performing computation in parallel, especially if there is no need in the complement operation. If the complement operation is present, then the total problem can be broken down into a set of subproblems, each covered by the corresponding complement operation, and which then are delegated for separate execution in parallel.

For example, the problem below could be processed in parallel as follows:

$$\langle\overline{x \mid 1\rangle\langle 0 \mid y\rangle\langle\overline{x \mid o\rangle\langle 1 \mid y}}\rangle\langle x \mid o\rangle\langle\overline{z \mid y\rangle\langle 1 \mid z}\rangle$$

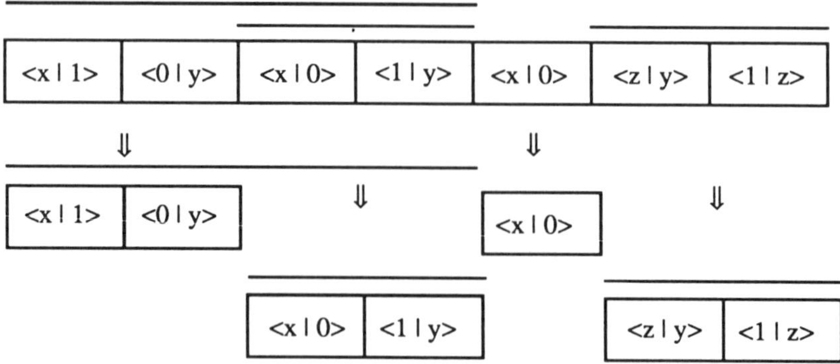

Guided by the complement operation the problem is first divided into four parallel subproblems and then the results of all parallel computations are brought together to complete evaluation.

Though the methods explained here allow the introduction of parallel processing, it remains globally a sequential approach in the sense that there are no systematic links between the operations executed in different logic subproblems. Genuinely parallel logic processing not only must allow several logic operators to be handled simultaneously but there must also be definite rules that relate the operators involved, very much in the same manner as the results of human brain operations arise not only from massively parallel computations, but computations which also bonded among themselves in an organized and congruent fashion.

Parallel logic architertures

With the limits of processing speed and microminiaturization already in sight, it has become apparent in recent years that any major leaps in computer power will have to involve fundamental changes in the design of processor and computer architecture. The conventional computer is based on the serial Von Neumann architecture: it has one central processor unit, one memory bank, and a communication channel linking them. Instructions from the processor and data from the memory flow back and forth through the communication channel, which limits processing speed, the well-known bottleneck problem.

In order to resolve the bottlenecks, parallel architecture may have many processors and memory banks. Supercomputers are made of several Von Neumann processors, which can work together on a problem to solve it faster than a single processor. Each processor can work alone or in parallel with other processors. There is essentially only one way to construct a serial machine, but infinitely many ways to build a parallel one. For once we open the door to

morc than one processor, we admit numerous possibilities for connecting processors.
The real and non-trivial problem consists in interconnecting processors and controlling their operations. High expectations of the use of parallelism in the interconnection of processing and memory units gave rise to a number of different parallel architectures, each with their own advantages and limitations.
Inspired by the massive parallelism of the brain, which performs millions of operations simultaneously, the ultimate goal is decentralized architecture, made up of comparatively simple, but very large number of processors.
Decentralized architecture may turn out to be useful for making logical inferences on large knowledge bases. But how to make data always go where it is supposed to go and at the right time? The problem of coordination remains the main problem in parallel computing. An exceptional advantage of multiplicative circuitry resides in the possibility of expanding computer power that is both simple and uniform. Since the machine is designed homogeneously from uniform multiplicative gates <x | L | y>, their number can be easily increased if necessary.
One of the basic parallel architectures is the square processor array, schematically presented below:

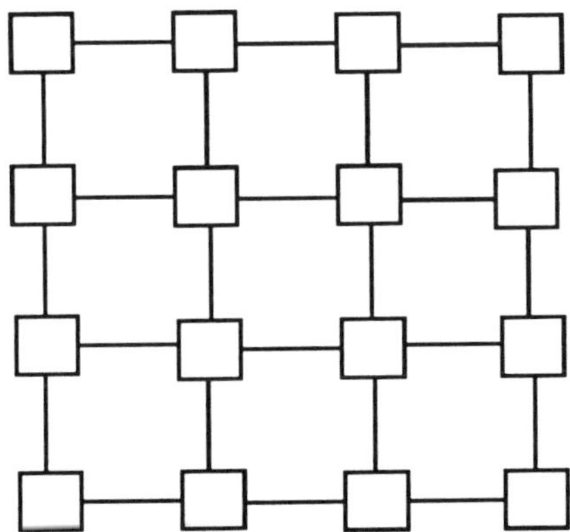

From an algorithmic point of view, the square array is the most natural for any problem involving matrices. Matrix-vector and matrix-matrix products can be easily implemented on the square processor array.This is quite evident for special-purpose computational systems, and the chief issue, consequently, is not special but general-purpose processor machines, which can process in parallel arbitrary logic operations.

In the next section we examine parallel logic computation with the support of group theory, with the aim of designing a matrix logic processor whose operations can be described as a group.

5.2 Hypercube group processors

Clearly, in order to address the problem of global parallelism we must consider hypercube logic. Linked to the group-theoretic method it allows a number of important questions to be answered concerning the design of parallel matrix processors and sets definite bounds on their performance. The unification of group-theoretic and logic notions which follows not only provides a rigorous technical guide for the design of parallel processors, it also has general theoretical value, inspiring a development of group logic.

As we recall, the hypercube operators are (n × n) logic operators, which can be partitioned after a finite number of steps into the basic (2 × 2) L-operators. Since the results presented here are applicable to a hypercube logic of any order, it will be sufficient for our purposes if we consider (4 × 4) hypercube operators of the form

$$H = \begin{pmatrix} L_{11} & L_{12} \\ L_{21} & L_{22} \end{pmatrix}$$

where L_{ij}, i,j = 1,2 are some (2 × 2) logic operators.
Such hypercube operators can be realized by the sets of 16 binary switches, each capable of being in one of two possible states described by 1 and 0:

output					
← y_1	0	1	0	0	←
← y_2	0	0	1	0	←
← y_3	1	0	0	0	←
← y_4	0	0	0	1	←

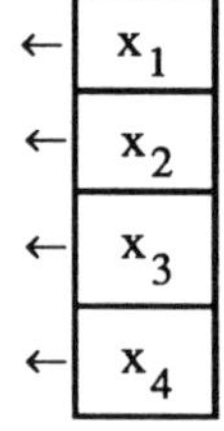

output input

As its input the matrix processor receives 4-dimensional logic vectors (bra or ket)

$$(x_1 \; x_2 \; x_3 \; x_4)$$

and outputs 4-dimensional logic vectors (bra or ket)

$$(y_1 \; y_2 \; y_2 \; y_3) ,$$

The mappings of input vectors onto output vectors are uniquely defined, with components y_1, y_2, y_3, y_4 being simply a new combinatoric of the components x_1, x_2, x_3, x_4. In the 4-dimensional logic space there will be 4! = 24 different mappings of this sort. We can present these in the following graphs, where arrows indicate the change-over between different components of the logic vectors:

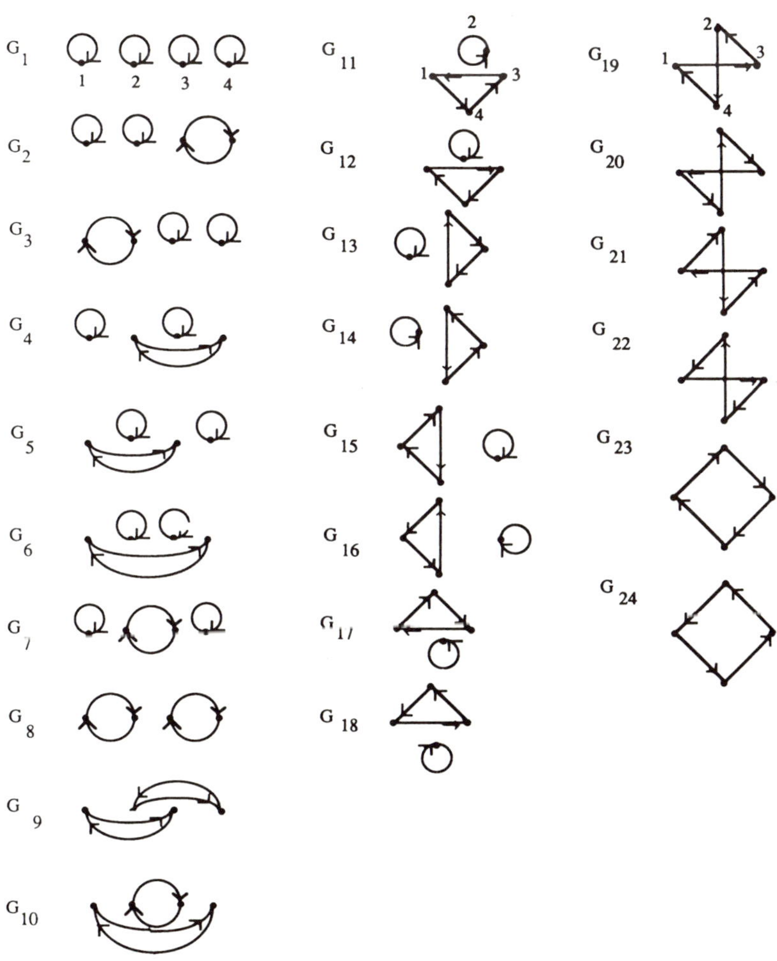

The graphs presented above exhaust all possible mappings of logic vectors which can be achieved on a (4 × 4) matrix processor. Higher-order processors allow greater flexibility to be achieved; in the general case, for a processor of order (n × n) there will be n! possible mappings.

Our task is now to determine the hypercube logic operators H_i which can uniquely perform the mapping given above; in so doing we will uniquely define the operations of the matrix processor.

The problem can be conveniently formulated as a Black Box problem, where we are given the set of input vectors and the corresponding set of output vectors, and the task is to determine the hypercube logic operators which execute the desired mappings:

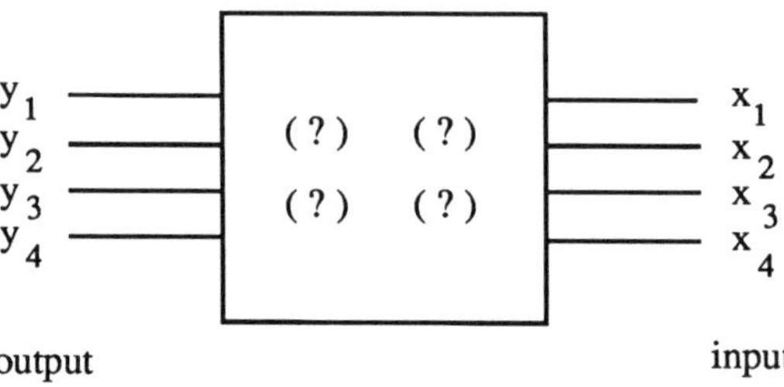

The question marks indicate unknown L-operators which we would like to define in unique fashion, thus giving each hypercube operator a unique representation. After a number of trials with the matrix processor, presented in the form of the Black Box, we arrive at the following set of H-operators which gives an unique solution to the problem:

$$H_1 = \begin{pmatrix} \leftrightarrow & 0 \\ 0 & \leftrightarrow \end{pmatrix} \qquad H_{13} = \begin{pmatrix} \uparrow & \backslash \\ \wedge & / \end{pmatrix}$$

$$H_2 = \begin{pmatrix} \leftrightarrow & 0 \\ 0 & \rceil \end{pmatrix} \qquad H_{14} = \begin{pmatrix} \uparrow & \wedge \\ / & \backslash \end{pmatrix}$$

$$H_3 = \begin{pmatrix} \rceil & 0 \\ 0 & \leftrightarrow \end{pmatrix} \qquad H_{15} = \begin{pmatrix} / & \wedge \\ \backslash & \uparrow \end{pmatrix}$$

$$H_4 = \begin{pmatrix} \uparrow & \wedge \\ \wedge & \uparrow \end{pmatrix} \qquad H_{16} = \begin{pmatrix} \backslash & / \\ \wedge & \uparrow \end{pmatrix}$$

$$H_5 = \begin{pmatrix} \wedge & \uparrow \\ \uparrow & \wedge \end{pmatrix} \qquad H_{17} = \begin{pmatrix} / & \backslash \\ \uparrow & \wedge \end{pmatrix}$$

$$H_6 = \begin{pmatrix} \wedge & / \\ \backslash & \uparrow \end{pmatrix} \qquad H_{18} = \begin{pmatrix} \backslash & \uparrow \\ / & \wedge \end{pmatrix}$$

$$H_7 = \begin{pmatrix} \uparrow & \backslash \\ / & \wedge \end{pmatrix} \qquad H_{19} = \begin{pmatrix} 0 & \leftrightarrow \\ \rceil & 0 \end{pmatrix}$$

$$H_8 = \begin{pmatrix} \rceil & 0 \\ 0 & \rceil \end{pmatrix} \qquad H_{20} = \begin{pmatrix} 0 & \rceil \\ \leftrightarrow & 0 \end{pmatrix}$$

$$H_9 = \begin{pmatrix} 0 & \leftrightarrow \\ \leftrightarrow & 0 \end{pmatrix} \qquad H_{21} = \begin{pmatrix} / & \wedge \\ \uparrow & \backslash \end{pmatrix}$$

$$H_{10} = \begin{pmatrix} 0 & \rceil \\ \rceil & 0 \end{pmatrix} \qquad H_{22} = \begin{pmatrix} \backslash & \uparrow \\ \wedge & / \end{pmatrix}$$

$$H_{11} = \begin{pmatrix} \wedge & / \\ \uparrow & \backslash \end{pmatrix} \qquad H_{23} = \begin{pmatrix} / & \backslash \\ \backslash & / \end{pmatrix}$$

$$H_{12} = \begin{pmatrix} \wedge & \uparrow \\ \backslash & / \end{pmatrix} \qquad H_{24} = \begin{pmatrix} \backslash & / \\ / & \backslash \end{pmatrix}$$

The set of hypercube logic operators realized by the matrix processor is remarkable in several important respects:

1) hypercube operators are Cartesian sums of the L-operators of the weakly normalized adaptive logic $\Omega^{\varepsilon}_{weak}$:

$$\{ \leftrightarrow, \rceil, \wedge, /, \backslash, \uparrow, \hat{0} \}$$

2) the product of any two hypercube operators is again a hypercube operator;

3) for any hypercube operator H there exists a finite power a, such that H^a converges to the identity operator;

$$H^a = I .$$

We have the following series:

$$H_1^1 = I$$

$$H_2^2 = H_3^2 = H_4^2 = H_5^2 = H_6^2 = H_7^2 = H_8^2 = H_9^2 = H_{10}^2 = I$$

$$H_{11}^3 = H_{12}^3 = H_{13}^3 = H_{14}^3 = H_{15}^3 = H_{16}^3 = H_{17}^3 = H_{18}^3 = I$$

$$H_{19}^4 = H_{20}^4 = H_{21}^4 = H_{22}^4 = H_{23}^4 = H_{24}^4 = I$$

The existence of such convergences to the identity operator implies that for any hypercube operator there exists an inverse such that

$$H_i H_i^{-1} = H_i^{-1} H_i = I , \quad i = 1,24 .$$

Taking into account the associativity of matrix multiplication, we conclude that the matrix processor realizes a group operation, and that the totality of these operations forms a group.

Recall that a set G consisting of elements a,b,c,...

$$G = a,\ b,\ c,\ ...$$

is a group if the following conditions are satisfied:

CLOSURE

1. For any two elements $a,b \in G$, a rule of composition is defined

$$a * b = c \qquad \text{with } c \in G,$$

ASSOCIATIVITY

2. For composition, the associative law holds, that is, for $a,b,c \in G$

$$a * (b * c) = (a * b) * c,$$

IDENTITY

3. G contains a unit element I such that for every $a \in G$

$$a * I = I * a = a ,$$

INVERSE

4. For each $a \in G$ there exists the inverse element a^{-1} such that

$$a^{-1} * a = a * a^{-1} = I ,$$

COMMUTATION

5. The commutation law does not necessarily hold. If $a * b = b * a$, G is a commutative or abelian group, otherwise it is not commutative.

Comparing the properties of a group with the properties of the matrix logic processor discussed above leaves no doubt the latter functions as a group, with all the important implications which follow from that.

Identity transformation

The hypercube group matrix processor is a parallel logic reversible machine, in the sense that once it has reached the final step of computation, it can be played backwards and will retrace the steps of computation to the initial state. The execution of each group operation induces a transformation in the machine, for which there always exists another group operation such that if the two are performed one after the other the machine eventually returns to the initial state. In other words the identity operation is everywhere defined, and every pair of adjacent operations which are inverses of each other cancels the transformation.

The capability of retracing computational steps is a significant advantage of hypercube logic operators in comparison with (2×2) logic operators for which the inverse operations, and consequently, the identity transformation, cannot be universally defined: singular L-operators do not have inverses. Furthermore, with the exception of the equivalence operator $\leftrightarrow$ and the negation operator $\urcorner$, the computation of other inverses leads one out of the domain of positive logic.
Nothing of the kind is permitted to occur in hypercube logic, since otherwise the fundamental requirement of a group cannot be satisfied. With matrix multiplication as a group composition operation, the hypercube processor realizes multiplicative computations and is capable of realizing with any given operator its inverse, and consequently the identity transformation.

Orthogonality

In any fruitful generalization of matrix logic it is desirable to retain the useful features of the basic (2 × 2) matrix logic. Given that the fundamental basis set of (2 × 2) logic consists of the orthogonal logic operators {↔,⌉}, we define, in this restrictive sense, (2 × 2) logic as an orthogonal logic. Consequently, it seems reasonable to retain orthogonality in logic spaces of any dimension. The orthogonality condition restricts the set of possible L-operators which can be used for the design of hypercube H-operators to the set of adaptive logic:

$$\{ \leftrightarrow, ⌉, \wedge, /, \backslash, \uparrow, \hat{0} \}.$$

Though H-operators are composed from only 7 L-operators their number grows rapidly if the dimension of the logic vector space increases. In 8-dimensional vector space we can construct 40320 different hypercube operators.

The hypercube operators are orthogonal operators and induce automorphisms on the set of the given logic vectors. The orthogonality of the hypercube operators implies that they are represented by orthogonal matrices, with the transpose being equal to the inverse:

$$\tilde{H}_i = H_i^{-1}, \quad i = 1{,}24.$$

We have the following relation for the elements of the hypercube logic group. The operators H_1, H_2, H_3, H_4, H_5, H_6, H_7, H_8, H_9, H_{10}, as symmetric operators, are invariant to transposition, and consequently are inverses of themselves. The remaining 14 operators are divided into two subsets which are inverses of each other:

H	$H^{-1} = \tilde{H}$
H_{11}	$H_{11}^{-1} = H_{12}$
H_{13}	$H_{13}^{-1} = H_{14}$
H_{15}	$H_{15}^{-1} = H_{16}$
H_{17}	$H_{17}^{-1} = H_{18}$
H_{19}	$H_{19}^{-1} = H_{20}$
H_{21}	$H_{21}^{-1} = H_{22}$
H_{23}	$H_{23}^{-1} = H_{24}$

Due to the associativity of matrix multiplication, the existence of unit and inverse elements, we concluded that the hypercube operations of a matrix processor form a group which, given the orthogonality of its elements, is the group of orthogonal transformations.

Inverse transformation

The inverse transformation occupies a prominent place in computer applications. Frequently, when we are given some mapping for a logic function we are interested in determining an inverse mapping. For example, knowing that a given logic function evaluates to true, one would like to determine a corresponding assignment for input variables.
Often the issue arises whether to use rules for forward or backward inference, in other words, whether deduction should be driven by putting questions or by the available facts. In testing the integrated circuits that perform complex logic functions considerable effort is required to solve inverse problems.

Within the framework of conventional logic, there is no general algorithm for the solution of the inverse problem, since the inverse operation is not universally defined. Because hypercube logic operations are group operations, hypercube operators always have inverses, which supplies a simple and effective means for dealing with inverse transformations. Moreover, hypercube logic not only gives a simple resolution to the inverse problem, but also provides a generally applicable rule.

Consider a sequence of operations which, being applied to input logic vector $|x\rangle$, delivers some output logic vector $|y\rangle$:

$$|y\rangle = H_1 \; H_2 \; H_3 \dots H_n \, | \, x\rangle \, .$$

We can easily solve the inverse problem by applying the transposed sequence of operations as follows:

$$\widetilde{H_1 \; H_2 \; H_3 \dots H_n} \, | \, y\rangle = \widetilde{H_1 \; H_2 \; H_3 \dots H_n} \; H_1 \; H_2 \; H_3 \dots H_n \, | \, x\rangle$$

The inverse mapping is thus

$$|x\rangle = \tilde{H}_n \dots \tilde{H}_3 \; \tilde{H}_2 \; \tilde{H}_1 \, | \, y\rangle \, .$$

For the solution of the inverse problem very effective use can be made of the left and right logics, which notions were introduced in the preceding chapter. With the help of the left and right logics the inverse problem is reduced simply to a I/0 problem. Consider a transformation described by some hypercube operator H:

$$\langle x | H = \langle y | \, .$$

If we are concerned only with the inverse mapping of logic vectors then the inverse transformation is found in the adjacent vector space simply as

$$|x\rangle = H | y\rangle \, .$$

Indeed,

$$\tilde{H} | x\rangle = |y\rangle \, ,$$

from which follows:

$$H\tilde{H} | x\rangle = H | y\rangle$$

where

$$H\tilde{H} = I \, .$$

In the hypercube matrix processor the solution of the inverse problem is achieved by a simple interchange of transposed inputs and outputs, with no modification of a hypercube operator being required. Simply by feeding back the transposed output through the same machine we obtain its counter image. The property of the matrix processor discussed here may be very useful for the organization of the retrieval of information.

Memory organization

In great contrast to the brain, where there is no separation between memory and processing elements, and information can be stored in the connections of a distributed network, in Von Neumann computer architecture, memory and processing units are separated. There is strong evidence that this makes the computer awkward for certain intelligent tasks by comparison with the network of the brain.

Since each hypercube operator can be associated with a certain network structure, the group operations of the matrix processor can be characterized by a dual property.
On one hand, they represent the processing operators; on the other hand, they uniquely define the state and evolution of the memory. Thus, results of computation belong to the group both as units of memory and as units of processing. The activation of some hypercube logic operator is at the same time the actualization of specific memory loops. Changing the state of the memory will modify accordingly the result of processing and vice versa; and multiplication of two hypercube operators not only produces a new operator, it also at the same time forms a new configuration of the network and thus a new state of the memory.

An inversion of the orientation of the arrows in a given memory network is equivalent to the transposition of the corresponding hypercube operator, with the latter initiating an inverse operation. Thus a transformation and its tranpose can be put in correspondence with the storing and retrieval of information. Consider as an illustration:

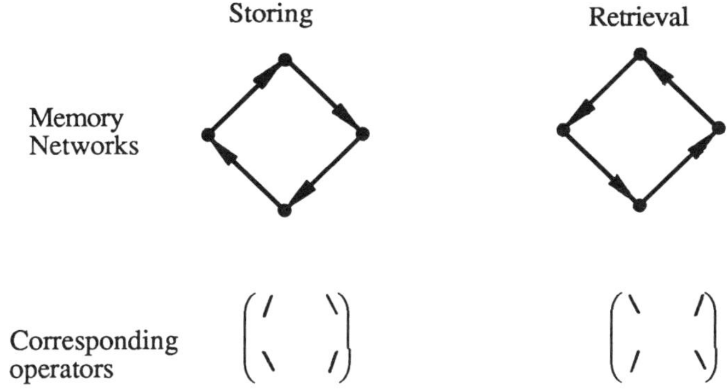

It is important to bear in mind that arrow inversion influences only non-symmetric hypercube operators, while symmetric operators and their corresponding memory networks are invariant under inversion.
Knowledge of a hypercube operator automatically gives us explicit information on the state of the memory. Conversely, a given memory pattern controls the structure of the processing logic operator, and, vice versa, a given logic operator determines a certain pattern of memory loops. In the view that the execution of a program on a matrix processor involves consecutive interactions of operators and their inverses together with identity, it is possible to evaluate some logic functions and simultaneously to consider their mapping onto memory, which includes the storing and retrieval of information.

Pure states

Suppose that the logic vectors received and outputted by the matrix processor are in a pure state. This implies that the excitation of one channel must be accompanied by the absence of excitation in the remainig channels. Then the inputs and outputs can be represented by hyper-Boolean logic vectors, that is n-dimensional normalized vectors $(x_1, x_2, \ldots x_n)$ in which one component is equal to 1 and the rest are 0's:

$$\sum_{i=1}^{n} x_i = 1 \; ; \qquad x_i = \begin{cases} 1 & - 1 \text{ component} \\ 0 & - n{-}1 \text{ components} \end{cases}$$

The task of the matrix processor is to translate a hyper-Boolean vector into another hyper-Boolean vector, which apparently means a shifting of 1 from a given position to some other position, including the same position as well. Operations on hyper-Boolean vectors are attractive from the computational point of view and give more room for flexibility. The desired transformation of a vector can be accomplished by several hypercube operators. For example, the bra vector (0, 1, 0, 0) can be translated into the bra vector (1, 0, 0, 0) by six hypercube operators: H_3, H_8, H_{16}, H_{18}, H_{22}, H_{24}. The bilinear products for hypercube operators are defined in a way identical to (2 × 2) logic operators, yet with new interesting properties:

$$\langle x \mid H \mid y \rangle = \begin{cases} 1 & - n \text{ cases} \\ 0 & - n(n-1) \text{ cases} \end{cases}$$

with

$$\langle x \mid H\tilde{H} \mid y \rangle = \langle x \mid y \rangle$$

$$\langle x \mid H \mid y \rangle = \langle y \mid H^{-1} \mid x \rangle$$

where $\langle x|$ and $|y\rangle$ are hyper-Boolean logic vectors.

Hypercube logic is richer in combinatorial possibilities than (2 × 2) logic. The dimensionality of a logic vector space determines the number of different cases in which a hypercube logic function of two arguments in pure states evaluates to 1. In 4-dimensional logic vector space, bilinear products in four cases yield the truth-value 1, and in twelve cases the truth-value 0.

Multispinor matrix processor

Since hyper-Boolean logic vectors have a single 1 and n – 1 zero elements, the appearance of several 1's would be a clear indication of an error. On the other hand, a shift of 1 from a correct to a wrong position may not be easily recognizable. In quest of higher fault tolerance operations, we may consider a multispinor multiplicative processor. The logic vectors are now defined in the n/2-out-of-n code, implying that each logic vector contains the same number of 1's and 0's.

Furthermore, all positions are subdivided into coupled spin pairs: (0, 1) or (1, 0), which are the simplest possible logic vectors. A multispinor logic vector may be written, for example, as follows:

$$\langle x| = (0\ 1 \quad 1\ 0 \quad 1\ 0 \quad 0\ 1)\,.$$

With the normalizing multiplier $(n/2)^{-1}$ we retain the essential property of a logic vector

$$\|x\| = (n/2)^{-1}\langle x \mid x\rangle = 1\,,$$

where n is the dimension of the given logic space.

Operations in such a code have high reliability since faulty signals will immediately lead to the destruction of the n/2-out-of-n code. The requirement of the preservation of the code can be satisfied by introducing cellular structure for multispinor operators. We employ the formalism of hypercube logic with each hypercube operator providing a cellular form into which a certain number of multispinor operators may be inserted. The insertion is done by replacing 0's with the operators $\hat{0}$, and 1's by the operators of the binary spinor bases:

$$\{\leftrightarrow, \urcorner, \perp, \top\} \quad \text{– left spinor base}$$
$$\{\leftrightarrow, \urcorner, \dashv, \vdash\} \quad \text{– right spinor base.}$$

The operators of spinor bases are the only logic operators which allow preservation of the n/2-out-of-n code, although this now applies separately to the left and right logics. Just as the product of two hypercube operators is again a hypercube operator, so also with multispinor operators. However, now we consider cellular inverses and cellular identity:

$$H_{cube}\,\tilde{H}_{cube} = I$$

$$H_{spin}\,\overset{\boxed{\sim}}{H}_{spin} = \boxed{I}$$

where $\boxed{\sim}$ denote cellular transposition, by which we transpose a cell against the principal diagonal and leave an operator inside a cell unchanged; $\boxed{I}$ is a cellular identity or an operator where the cellular main diagonal does not contain empty operators $\hat{0}$, whereas all other cells are empty. An example of cellular identity is given below:

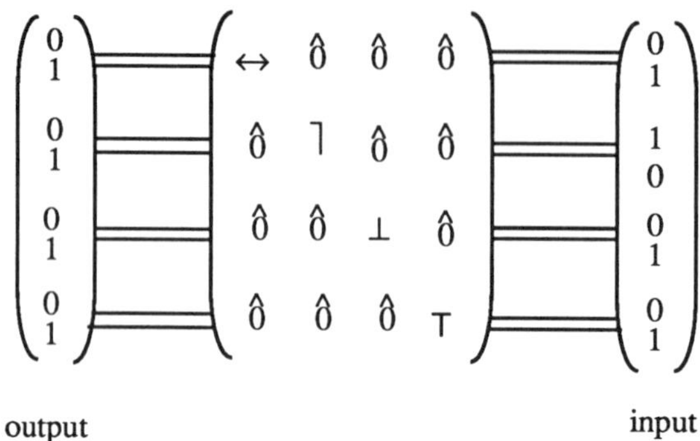

With respect to memory organization the difference between hypercube and multispinor processors consists in the fact that the knots of the memory network in the hypercube operator are not allowed to have more than one input and one output, whereas the multispinor operator may have multiple duplicate inputs or outputs.

Both processors offer the possibility of performing logic operations in parallel, and can be used for the design of general-purpose group machines, which are constructed by joining many finite state machines together.

5.3 Abelian and non-abelian machines

Hitherto we have investigated the matrix logic processor from the group-theoretic point of view. The group-theoretic approach was chosen for a number of important reasons. First of all we can conceive and implement global logic operations which are meaningful, since the inverse operation is everywhere defined.

In contrast to ordinary L-operators, some of which are singular operators without inverses, the inverses of hypercube operators are always defined, and can be computed in two ways: by using either the operation of matrix transposition or the approporiate power of a matrix operator: H^a. Secondly, the parallel group logic of a matrix processor is a Cartesian sum of the operators of the adaptive logic. Acting in parallel within a hypercube operator, these exhibit group properties.

The fact that group logic incorporates in itself the adaptive logic is a factor which deserves special attention.

Most logic problem solvers are forced to reason with inconsistent or incomplete information. This may require nonmonotonic reasoning that allows facts to change adaptively rather than only allowing them to be added. Frequently, problem solvers are confronted with the need to select among plausible alternatives or to draw adaptively plausible inferences from incomplete data. Thus, virtually all practical problem solvers must incorporate an adaptive mechanism to manage this inconsistency.

In the matrix processor the adaptation capability is resolved automatically thanks to the group-theoretic requirements imposed on the hypercube logic operators. Whatever global logic operations are executed on the matrix processor, locally they remain within the domain of adaptive logic. Furthermore,given the appropriate degree of control, we can introduce local modifications into a hypercube operator, which are irrelevant for the overall result of computation.

Determining that the operations of a matrix processor are group logic operations, we have discussed their properties of associativity, closure, and the inverse and identity operations. One important aspect was thus far left out of our discussion: commutativity. In order to make this group-theoretic examination complete we shall examine the question of the commutativity of matrix processor operations.

Groups are usually divided into two main classes: commutative or abelian groups and non-commutative or non-abelian groups, depending on whether the commutation law holds or not. Similarly, matrix processors can be divided into abelian and non-abelian group machines.

The bulk of our discussion has been devoted to the investigation of (4×4) hypercube operators, which set in parallel two (2×2) adaptive logics. This was because their group properties are representative of the whole class of orthogonal group machines, where $n \geq 4$. From the point of view of applications it is mainly processors capable of accommodating large n that are of interest. However, orthogonality is not an exclusive property of H-operators. In the domain of (2×2) logic the operators of the orthogonal basis set $\{\leftrightarrow, \urcorner\} \cap \{x, +, -, 0, 1\}$, equivalence and negation:

$$\begin{pmatrix} 1 & 0 \\ 0 & 1 \end{pmatrix}, \qquad \begin{pmatrix} 0 & 1 \\ 1 & 0 \end{pmatrix},$$

are orthogonal operators. Apparently, the two operators of the orthogonal basis set can be considered as the simplest possible hypercube operators, which set in parallel two (1×1) scalar logics. These two operators form the rotation group of symmetry for which the commutation law holds, and consequently, the matrix processor resulting from such orthogonal logic will be an abelian group machine.

This fact is particularly important because it is the only possible abelian logic processor, with hyper-Boolean logic vectors being reduced to the basic logic vectors (1,0) and (0,1) as their most simple and minimal case.

Except for the orthogonal operators of the fundamental basis set, no other hypercube operators can generate a fully abelian commutative machine. Thus using higher-order logic operators, we are capable of retaining orthogonality but lose commutativity in globally parallel logic operations. This observation is significant not only from the applied point of view, but also in the more general theoretical framework of the intelligent capabilities of a computer.

In classical Turing theory the characterization of a computer is accomplished through the study of its input. In defining a parallel logic computer as a group, we depart from the classical theory and define the machine as a group with intrinsic built-in control over its own operations.

The elements of the group can be considered as the programs of the machine. The hypercube group is ideally suited for a number of reasons, most importantly because it is non-commutative, as one should expect any representation of a program to be.

Within the framework of the group approach we can speak of the programs as the group in which we define a unary operation, the inverse, and an associative binary operation, the product.

Group logic provides an effective means of control over the operations of the machine. Because the associative laws holds, we define the products as a string of hypercube operators or what we obtain from it after cancelling out all pairs of adjacent operators which are inverses of each other.

Apparently a matrix processor is an associative non-abelian machine with intrinsic group properties where any result of computation is again an element of a group. The group-theoretic approach allows some functions of control to be transferred to a processor, making it a more autonomous logic system. Imposing the condition of orthogonality we do not only oblige any result of computation to be again an element of a group but also predetermine the latter as being composed of logic operators of the adaptive logic base.

An understanding of the group character of parallel logic computations is important for the study of the intelligent capabilities of a computer.

5.4 Optical computing

The important issue in the construction of multiplicative matrix processors is the availability of appropriate hardware. Conventional technology provides a wide choice for engineering realizations. However, the development of matrix processors becomes especially attractive within the framework of the newly emerging optical technology.

Optical computing is currently exploring the frontiers of computing. The goal is to devise a machine that is better at handling classes of problems that do not lend themselves well to electronics or to develop an optical supercomputer that will outstrip even the fastest electronic supercomputers.

However, in spite of the impressive advance of optical hardware, it remains far behind traditional computers in the realization of logic. Though excellent for superfast numerical computations, optical computing devices are as yet no match for the conventional general-purpose computer, as far as system logic and programming are concerned.

In this respect the implementation of matrix logic in optical environments could be especially beneficial, serving as a blueprint for the development of the logic capabilities of a general-purpose optical computer. Furthermore, optical computing, as a multidisciplinary field, seeks contributions from computer science as well as from physics. And matrix logic, with its links both to physics and computer science, may prove to be a useful asset here.

The general type of operation performed by the discrete optical processor is a matrix-vector product of the form

$$|y\rangle = A\,|x\rangle\,,$$

where $|x\rangle$ is an m-dimensional input vector, $|y\rangle$ is an n-dimensional output vector, and A is an $(n \times m)$ matrix.

Elements of the input vector $|x\rangle$ and the matrix A are represented by light intensities, implying that they are non-negative and real. Negative or complex-valued vectors and matrices can be handled by representing their elements in terms of several non-negative and real components.

The elements of the input vector $|x\rangle$ are entered into the system in time sequence as pulses with intensities proportional to the desired values. The light from each pulse diverges and falls on the entire column of the matrix mask. After m pulses and charge transfers, the total charge accumulated on the ith element of the detector is precisely the ith element of the desired output vector $|y\rangle$.

In principle, all n elements of the ouput vector are available in parallel. In this case the elements of the input vector are introduced in parallel, and the elements of the output vector are detected in parallel on an array of independent detector elements. The light from each input element spreads into a vertical column, which illuminates a single column of the matrix mask, whereas each photodetector receives a light intensity proportional to the sum of the light intensities transmitted by an entire row of the matrix mask.

In recent years, various innovative ideas have surfaced for using the parallel matrix-vector multiplier in an iterative mode. Such systems are utilized for solving sets of simultaneous linear equations of the form

$$|y\rangle = L\,|\,x\rangle$$

where L is a known matrix and $|y\rangle$ is a known vector, and the goal is to determine the unknown vector $|x\rangle$.

A feedback arrangement between detector array and source array can be shown to result, after a sufficient number of iterations, in convergence of the output vector to the desired solution $|x\rangle$.

Optical implementations of iterative methods have been proposed for finding eigenvalues and eigenvectors of matrices. A more recent application of iterative processing using a parallel matrix-vector multiplier has been applied to the construction of associative memories, a subject that is currently receiving much attention. The matrix mask contains a series of row vectors which are to be stored in the memory.

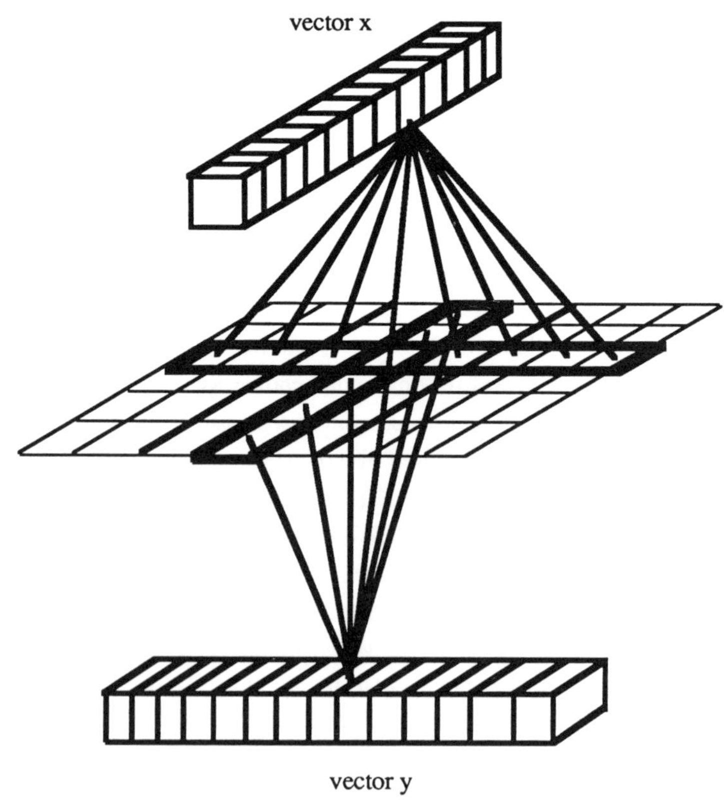

vector y

Parallel matrix logic processor

At the input, a part of one of the stored vectors is applied as the input vector, and the output vector is fed back to the input. After some iterations, the output vector converges to the stored vector that most closely matched the original input vector.

Discrete optical processors have been considered for very large system integration implementation.

There are in fact many options for optical architecture, depending in part on the operation to be performed. Our key interest concerns a processor designed to perform a matrix-vector product, followed by computation of an inner product. Thus the central processing unit functions as a familiar basic device <x | L | y>, the geometry of which is identical to that of a matrix-vector multiplier.

An attack vector <x| is represented by a string of light spots, generated by a linear array of laser diodes. The light from each diode is spread to illuminate all the values in the corresponding column in the matrix L, and thus is multiplied by each value. The values from each row in the matrix L is then focused onto each detector in the defence linear array |y> behind the spatial light modulator.

This matrix processor provides a method of connecting all possible attacks (inputs) with all possible defences (outputs), where each laser illuminates an entire column of switches and each detector receives light from the entire row.

5.5 Intelligent capabilities

This concluding section is included not only because of its intrinsic interest, but also, and more importantly, to emphasize the importance of matrix logic for modelling and understanding intelligence, both natural and artificial, computationally.

The initial interest of computer designers in processing speed and device miniaturization is currently being increasingly replaced by interest in the intelligent capabilities of the computer. Artificial intelligence seeks to expand the boundaries of computer science in order to solve problems traditionally considered to require human intelligence. The ultimate goal is to exploit wholly new computer architectures that will reason and behave more like a human brain than an ordinary number cruncher. However, in contrast to speed and degree of miniaturization, which are reaching the physical limits imposed by relativity and quantum physics, progress with developing intelligent capabilities has been less striking and remains largely at an embryonic stage. It is becoming apparent that neither parallelism nor sophisticated software engineering alone can provide an answer to artificial intelligence.

Though an existence proof is at hand, the nature of intelligence phenomena and their underlying logic are understood only vaguely. We do not fully understand the mechanics of how the brain does its operations, how it organizes, processes, stores and retrieves information. Then how could we endow computer with attributes we really cannot define? To come to grips with the problems that seem to defy not only conventional computer methods but also the techniques of the present artificial intelligence system, one needs a deeper exploration of logic.

The human brain displays a wide variety of intelligent functions, such as intuition, association, creativity, induction, learning and recognition. However, because it is not clear how these actually operate, there are no effective algorithms to achieve them in a computer. Among the intelligent functions only the algorithms of computation, and recently, deduction have been fully investigated, the former giving rise to the conventional Von Neumann computer, the latter to knowledge processing and the computer expert systems of the fifth generation. In spite of the intensive research which is being conducted into high-level intelligence, algorithms for other intelligent functions remain largely unknown.

In conventional computing the ratio between numerical computation or number crunching and logic operations is very heavily shifted towards number crunching, leaving little space for the implementation of logic-dependent intelligent capabilities, which leads to "knowledge bottlenecks" and reduces the effectiveness of parallel processing.

Whereas conventional computing excels at problems that lend themselves to be expressed in numerical terms, it has not been effective in dealing with problems that require high-level intelligence.

Research has determined that conventional programming languages which were developed for repetitive operations on numbers, are inefficient and inadequate wherever high-level intelligence and symbolic processing are involved.

Matrix logic offers viable solutions for dealing with this situation. The discovery of the connection between logic and matrix operations allows us to translate fully logic relations into the language of numbers appearing in algebraic equations. As a result, matrix logic offers new prospects for overcoming limitations in current knowledge processing technology, allowing one to express both number crunching and "logic crunching" in the uniform terms of computation. With its ability to utilize numerical operations in the logic domain and thus to incorporate in the same language pure numerical and symbolic logical operations, matrix logic promises to shed new light on the algorithms of hihg-level intelligence.

Since the algorithms of computation are known and have been fully studied, and since matrix logic has extended the full power of mathematical computation into the domain of logic, this offers the prospect of defining the algorithms of high-level intelligence in terms of computation.

Exploration of the group properties of intelligence is another important factor providing for accommodation of parallelism inherent in intelligent operations. Even more important for the field of AI is conclusion, reached in the preceding chapter, that the intelligence code is a tensor. An attempt to provide an answer to AI, without taking into account the tensor nature of high-level intelligence, is unlikely to be successful.

Autoproducts

Though the simplicity of the suggestion that reasoning is just computation may insult the complexity of the human mind, no valid scientific arguments to the contrary are known. Quite the opposite, the breakthrough in the development of machine intelligence very much depends on understanding intelligence computationally.
What is more, the exploration of the potential of matrix logic for the study of intelligence calls for a re-examination of such a fundamental issue as what we must understand by computing. Contemporary studies in artificial intelligence are all concerned with "scalar intelligence", failing to realize not only the matrix, but also and more importantly, the tensor nature of the intelligence code.

Another crucial reason why high-level intelligence resists formalization is related to the autonomous character of intelligent operations. In phenomenological terms this autonomous capability allows an intelligent system to postpone direct responses to stimuli from the environment and to carry out necessary evaluations prior to actions. This is the essential aspect of high-level intelligence which distinguishes it from ordinary data processing, information processing and knowledge processing.
Taking into account that the autonomous capability in mathematical terms implies a closed topology, one may reasonably incline to associate high-level intelligence with closed loop structures. Corresponding closed logic structures can be constructed from linear matrix logic expressions by connecting the last ket vector of an expression with its first bra vector. The proper writing of an expression in closed logic thus will require at least two-dimensional representations, with the actual expression taking the form of a circular string or autoproduct consisting of a finite number of bilinear products:

$$\overbrace{\langle x_1 | L_1 | y_1 \rangle \langle x_2 | L_2 | y_2 \rangle \langle x_3 | L_3 | y_3 \rangle \langle x_4 | L_4 | y_4 \rangle}^{x}$$

The exploration of the theoretical and engineering issues of closed logic leads to several interesting results. For linear logic in the inner product or fundamental product representations, the evaluation or reading of a linear expression can begin from any point and spread in either direction to the left or to the right or both, obeying, of course, the rule of the complement operator.

Such flexibility is due to the fact that the associative law of matrix multiplication holds. Because of the associativity of linear matrix logic, the result of computation is indifferent to the order of operations and is identical for all possible orders which do not violate the rule of the complement operator.

Being accustomed to the fact that the associativity of multiplication universally holds, one does not question whether it is indeed applicable with respect to autonomous intelligent operations.

A surprising novelty brought about by closed logic consists in its creating a new kind of matrix relation in which the associativity law cannot any longer be maintained. The result of computation comes to depend on which point the reading of an expression begins at, just as the proper decoding of genetic information requires a correct initialization point.

Furthermore, in contrast to linear logic, where the result of computation is a scalar value, in closed logic an expression can evaluate to either a scalar, which is the same should a closed string decay into a linear one, or to a set of logic operators, depending from which point the computation has begun. If the initialization point is a logic operator or ket logic vector, the result of computation is a logic operator, but should it be a bra logic vector, then the result is a scalar, identical for the evaluation of the corresponding linear expression.

Clearly, the autoproduct is a mathematical object with new fundamental properties. In closed logic the associativity law surrenders to a more general principle, according to which the mathematical nature of the product, surprisingly enough, becomes dependent on the order of operations!

We began this book with the definitions of the inner and outer products of two vectors. As now becomes apparent both these fundamental operations can be joined into the single concept of the minimal autoproduct which in its scalar mode produces the inner product, and in the operator mode, the outer product.

Below we consider the example of autoproducts formed as the closure of the contradiction:

$$\begin{array}{l} \qquad\quad 1 \;\; 2 \;\; 3 \qquad 4 \;\; 5 \;\; 6 \\ \alpha = \langle x \mid \wedge \mid y\rangle\langle x \mid \downarrow \mid y\rangle \equiv 0 \end{array}$$

and the corresponding tautology

$$\overline{\alpha} = \langle \overline{x \mid \wedge \mid y\rangle\langle x \mid \downarrow \mid} y\rangle \equiv 1$$

The numbering on the top indicates the start-positions for computation.

$\langle x\vert$	$\vert y\rangle$	start position	α	$\bar{\alpha}$	α^2	$\overline{\alpha^2}$
$\langle 0\vert$	$\vert 0\rangle$	1	$0\cdot 1=0$	1	0	1
$\langle 0\vert$	$\vert 1\rangle$	1	$0\cdot 1=0$	1	0	1
$\langle 1\vert$	$\vert 0\rangle$	1	$0\cdot 1=0$	1	0	1
$\langle 1\vert$	$\vert 1\rangle$	1	$1\cdot 0=0$	1	0	1
$\langle 0\vert$	$\vert 0\rangle$	2	$\wedge\uparrow\downarrow\uparrow=\hat{0}$	$\hat{\uparrow}$	$\hat{0}$	$\hat{\uparrow}$
$\langle 0\vert$	$\vert 1\rangle$	2	$\wedge\backslash\downarrow\backslash=\backslash$	$\rightarrow$	$\hat{0}$	$\hat{\uparrow}$
$\langle 1\vert$	$\vert 0\rangle$	2	$\wedge/\downarrow/=\hat{0}$	$\hat{\uparrow}$	$\hat{0}$	$\hat{\uparrow}$
$\langle 1\vert$	$\vert 1\rangle$	2	$\wedge\wedge\downarrow\wedge=\hat{0}$	$\hat{\uparrow}$	$\hat{0}$	$\hat{\uparrow}$
$\langle 0\vert$	$\vert 0\rangle$	3	$\uparrow\downarrow\uparrow\wedge=\hat{0}$	$\hat{\uparrow}$	$\hat{0}$	$\hat{\uparrow}$
$\langle 0\vert$	$\vert 1\rangle$	3	$\backslash\downarrow\backslash\wedge=\hat{0}$	$\hat{\uparrow}$	$\hat{0}$	$\hat{\uparrow}$
$\langle 1\vert$	$\vert 0\rangle$	3	$/\downarrow/\wedge=/$	$\leftarrow$	$\hat{0}$	$\hat{\uparrow}$
$\langle 1\vert$	$\vert 1\rangle$	3	$\wedge\downarrow\wedge\wedge=\hat{0}$	$\hat{\uparrow}$	$\hat{0}$	$\hat{\uparrow}$
$\langle 0\vert$	$\vert 0\rangle$	4	$1\cdot 0=0$	1	0	1
$\langle 0\vert$	$\vert 1\rangle$	4	$1\cdot 0=0$	1	0	1
$\langle 1\vert$	$\vert 0\rangle$	4	$1\cdot 0=0$	1	0	1
$\langle 1\vert$	$\vert 1\rangle$	4	$0\cdot 1=0$	1	0	1
$\langle 0\vert$	$\vert 0\rangle$	5	$\downarrow\uparrow\wedge\uparrow=\hat{0}$	$\hat{\uparrow}$	$\hat{0}$	$\hat{\uparrow}$
$\langle 0\vert$	$\vert 1\rangle$	5	$\downarrow\backslash\wedge\backslash=\hat{0}$	$\hat{\uparrow}$	$\hat{0}$	$\hat{\uparrow}$
$\langle 1\vert$	$\vert 0\rangle$	5	$\downarrow/\wedge/=\hat{0}$	$\hat{\uparrow}$	$\hat{0}$	$\hat{\uparrow}$
$\langle 1\vert$	$\vert 1\rangle$	5	$\downarrow\wedge\wedge\wedge=/$	$\leftarrow$	$\hat{0}$	$\hat{\uparrow}$
$\langle 0\vert$	$\vert 0\rangle$	6	$\uparrow\wedge\uparrow\downarrow=\hat{0}$	$\hat{\uparrow}$	$\hat{0}$	$\hat{\uparrow}$
$\langle 0\vert$	$\vert 1\rangle$	6	$\backslash\wedge\backslash\downarrow=\hat{0}$	$\hat{\uparrow}$	$\hat{0}$	$\hat{\uparrow}$
$\langle 1\vert$	$\vert 0\rangle$	6	$/\wedge/\downarrow=\hat{0}$	$\hat{\uparrow}$	$\hat{0}$	$\hat{\uparrow}$
$\langle 1\vert$	$\vert 1\rangle$	6	$\wedge\wedge\wedge\downarrow=\backslash$	$\rightarrow$	$\hat{0}$	$\hat{\uparrow}$

While in linear logic the fundamental product of contradiction always evaluates to 0, the corresponding autoproducts are uniform only for the scalar mode, but vary for the operator mode. If computation is halted after a single iteration, computing from different start-positions produces different operators. Steady states, $\hat{0}$, for contradiction, and $\hat{\uparrow}$, for tautology, however, can be achieved by the second iteration.

The autoproduct is clearly an object with unusual mathematical properties. In an attempt to answer the question what it actually represents, a scalar or an operator, we are confronted with a situation that is similar in some sense to the wave-particle dualism in quantum physics, where in one measurement condition a quantum process manifests particle properties, but in another measurement condition, wave properties.

Similarly, the autoproduct is neither a scalar nor an operator, but both a scalar and an operator depending on whether the internal or external mode of reasoning is required.
The interiorization and exteriorization processes lend themselves most naturally to treatment by autoproducts and together with the non-associativity of closed logic provide important clues to the puzzle of high-level intelligence.

With bilinear products as the minimal coding units, the duality principle of closed logic described above allows an intelligent system to realize a congruently two modes of operations: the scalar mode and the operator mode.
Since ideally, an intelligent machine must be able to accept both scalar and matrix signals, equal importance attaches to implementations of closed logic in computer science, where a formidable goal thus can be formulated as the design of a non-associative logic machine. In this context the development of machine intelligence is as much an engineering problem as it is a fundamental problem.

MATHEMATICAL RESUME

Diagonal principle

$\overline{\text{'A'}} = \neg\tilde{A}\neg$
$\exists = \neg\forall\neg$
$\Box = \neg\Diamond\neg$
$L^{*} = \neg\bar{L}\neg$

Matrix logic generators

$\{|x>, <y|\}$

Alphabet of truth-values

$\{0, 1, -1, i, \varepsilon\}$

0 – false
1 – true
– 1 – negative
i – imaginary
ε – approximate

Adaptive logic

$\{\leftrightarrow, \neg, \wedge, \backslash, /, \uparrow, \hat{0}\}$

Hypercube group logic

$<x | H | y> = <y | H^{-1} | x>$
$<x | H\tilde{H} | y> = <x | y>$

Spinor implication

$$\langle x \,|\, \circ\!\!\rightarrow \,|\, -x \rangle \langle -x \,|\, \circ\!\!\rightarrow \,|\, x \rangle = \langle x \,|\, \leftarrow\!\circ\!\rightarrow \,|\, x \rangle$$

Tensor logic products

$$\sum_{i}^{n} \bigwedge_{j}^{m} \langle e_k \,|\, L_j \,|\, e_r \rangle, \qquad L_j \in \Omega^{+} \cap \Omega^{-}$$

Autoproducts

$$\overset{x}{\overline{\langle x \,|\, y \rangle}} = \begin{cases} \langle x \,|\, y \rangle & \text{– scalar mode} \\ |y\rangle\langle x| & \text{– operator mode} \end{cases}$$

INDEX